ENGINEERING ENTREPRENEURSHIP FROM IDEA TO BUSINESS PLAN

This is a book for engineers and scientists who have the aptitude and education to create value through new products that could become income-producing businesses for them as well as for potential investors. The book uses short chapters without lengthy, distracting essays. The book is organized as a ten-week blueprint for value creation. Many first-time inventors going through the ten-week exercise covering the inventor's idea, patent application, and business model + business plan have become serial inventors of a wide range of products; they won multiple awards in business-plan contests and have attracted investment from wealthy investors, who see value in their technology-based innovations. Recent growth in business plan contests and angel investment clubs is evidence that many wealthy investors are seeking technology-based innovations in which to invest.

The book may be used by engineering students or engineers without a teacher. However, the book can be used in a college course as well, with teams made of three or four students per team. If you are a student or teacher, visit the book's website (www.engineer-entrepreneur-book.com) for useful supplementary materials.

Paul Swamidass, Thomas Walter Professor at Raymond J. Harbert College of Business, Auburn University, is a mechanical engineer with a doctoral degree in business management from the University of Washington, Seattle. He was a manufacturing manager before a long career in university teaching and research in the United States. At Auburn University, Swamidass was Director of the Thomas Walter Center for Technology Management from 2005 to 2014. He is the author of multiple books and 100+ scholarly publications.

During 2010–2013, Swamidass conducted four annual business start-up or business plan contests, open to all Auburn University students, for cash awards to help kick-start new student businesses.

As a pro se inventor-applicant, Swamidass was granted four US patents. His experience is described in the *John Marshall Review of Intellectual Property Law* (2010) paper and in an *Inventor's Digest* article aimed at individual inventors. In 2013, at the American Society for Engineering Education (ASEE) national conference in Atlanta, he conducted a workshop for engineering college professors on teaching invention and entrepreneurship to engineers.

Engineering Entrepreneurship
From Idea to Business Plan

A Guide for Innovative Engineers and Scientists

Paul Swamidass

Auburn University

CAMBRIDGE
UNIVERSITY PRESS

CAMBRIDGE
UNIVERSITY PRESS

University Printing House, Cambridge CB2 8BS, United Kingdom

One Liberty Plaza, 20th Floor, New York, NY 10006, USA

477 Williamstown Road, Port Melbourne, VIC 3207, Australia

314-321, 3rd Floor, Plot 3, Splendor Forum, Jasola District Centre, New Delhi - 110025, India

79 Anson Road, #06-04/06, Singapore 079906

Cambridge University Press is part of the University of Cambridge.

It furthers the University's mission by disseminating knowledge in the pursuit of
education, learning and research at the highest international levels of excellence.

www.cambridge.org
Information on this title: www.cambridge.org/9781107651647

First published 2016

A catalogue record for this publication is available from the British Library

Library of Congress Cataloging in Publication data
Names: Swamidass, Paul M., author.
Title: Engineering entrepreneurship from idea to business plan: a guide for
innovative engineers and scientists / Paul Swamidass, Auburn University.
Description: New York, NY, USA: Cambridge University Press, 2016. |
Includes bibliographical references and index.
Identifiers: LCCN 2016026640 | ISBN 9781107651647 (paper back)
Subjects: LCSH: Engineering – Vocational guidance. | New business enterprises. |
Entrepreneurship. | Business planning.
Classification: LCC TA157.S87 2016 | DDC 620.0068/1–dc23
LC record available at https://lccn.loc.gov/2016026640

ISBN 978-1-107-65164-7 Paperback

This is a book for all engineers and scientists brimming with new ideas.

This book is not a legal document. It covers the practices of the US Patents and Trademarks Office (USPTO) that permit an individual to apply for and get a US patent without the representation of patent attorneys; such applicants are called pro se inventor-applicants. The USPTO provides many useful and practical services to all individual pro se applicants during the application phase as well as the examination phase through their website, phone consultations, and through the examiner assigned to examine the Nonprovisional (Utility) patent applications (see MPEP 707.07(j)). A majority of individual inventors may be able to secure a patent in the United States without the services of a patent attorney if they learn how to. This book can serve those who cannot afford the services of a patent attorney but are willing to learn and work diligently to secure patents for their inventions; a few patent applications should give the inventor considerable practical knowledge and more confidence to invent frequently and apply as a pro se. However, if you need additional or advanced legal advice, and if you can afford it, seek a legal professional.

To my parents who modeled faith, hope and love (I Cor. 13:13) with a good measure of hard work, which is the secret sauce in all successful innovations.

Contents

PREPARING TO BE AN ENTREPRENEUR

Preface

This book is about value creation. Every day, engineers/scientists create value for their employers. The book is organized as a ten-week blueprint for value creation. It is designed to take a first-time inventor from an idea to his/her first patent application and business model + business plan in 10 weeks. Once inventors get the hang of it, they often enthusiastically turn into serial inventors of a wide range of products attractive to wealthy investors.

Thus, this book teaches readers how they can create value for themselves and for wealthy investors. Engineers/scientists have the education and training to invent and create new products, which could become income-producing businesses for inventors and wealthy investors. One graduate civil engineering student at my university, with the help of a workshop based on this book, developed two products and over two years won multiple awards for his products as well as his business plan, and received $180,000 in funding from investors several months after graduation in early 2016. Over a two-year period, he wisely obtained and used suggestions from several mentors. Two undergraduate students using the contents of this book in a class have formed a team to commercialize an invention that can take small loads of cargo into upper atmosphere more cost efficiently than other commercial options today. They have won awards for the invention and attracted interest from investors likely to invest more than a million dollars in their start-up business in 2016.

This book may be used by individual inventors, business plan contestants, or as a resource material in makerspaces, inventor clubs, or classrooms. The book uses short chapters and tries to get directly to the point without lengthy, distracting, mind-numbing essays; it is based on the assumption that bright engineers and scientists quickly understand the essential substance of the chapter and are anxious to move on.

This book covers a wide range of topics, from innovation, to finding creative ideas, to product development, to patent application drafting and filing, to business model development, and finally to business plan development with cash flow to help with valuation of the business. Toward the end of the book a few chapters offer a glimpse of what lies beyond the business plan, including chapters on teamwork and leadership, essential for business success. With such a range of topics between the covers, the depth of coverage in more than forty chapters of the book is adequate for a first-time inventor from the engineering/science field to get a complete picture of inventing, patenting and business start-ups – everything necessary to act. However, if the reader wants to know more about any given topic such as engineering product development, or marketing, etc., there are a number of outstanding books on the market that are devoted entirely to each topic.

There are numerous technology-based business plan contests with large monetary awards cropping up all over the United States, as well as in faraway countries such as India and China. The rapid growth in the number of business plan contests is a clear sign that many wealthy investors are looking for good technology-based businesses for investment. The TV show *Shark Tank*, premiering on ABC and with reruns on CNBC, is a good testament to this fact.

The book may be used with or without a teacher. The key features of this book enable readers to accomplish the following:

1. Sharpen their new idea
2. Turn an idea into a useful/commercial product
3. Conduct patent search and complete a Provisional and/or Utility patent application
4. Collect requisite data and prepare a business model
5. Collect additional data to prepare a business plan with five-year cash flow for valuation purposes to market the business to investors in exchange for equity, and to attract partners or company managers

A teacher could use the book in a college course with teams made of three or four students per team. The book is organized for a ten-week curriculum; however, I have used it in a fifteen-week semester quite frequently. The book enables teachers to pace the course over a ten-week or fifteen-week semester, even if they have never taught a similar course before; a course syllabus for the fifteen-week course can be downloaded from the website supplementing this book.[1] If you are a teacher, please visit the website and leave your suggestions and feedback. If you are a student visiting the website, please leave your feedback and share your success story as well as lessons learned.

I used a workbook of about forty pages with this book to enable students to plunge into prompt action after a short introduction to a chapter; information on the workbook is available through the same website.

The idea for the book was suggested by one of my students many years ago. Like all good innovations, it took shape slowly and, despite its current final shape, is still work in progress, subject to changes and improvements on the basis of the most current developments in engineering-associated industries and patent laws.

I sincerely thank Jay Clark, Haitham Eletrabi, David Mixson, and Hephzibah Stephen for sharing their valuable knowledge and experiences through the chapters contributed to this book.

---

[1] http://www.engineer-entrepreneur-book.com/

Acknowledgments

I gratefully acknowledge the generous gift of Mr. and Mrs. Tom Walter to the Thomas Walter Center for Technology Management, Auburn University, in 2006. Their large gift and other gifts earlier in Mr. Walter's honor by the Perot Foundation enabled the prolonged success of the Business-Engineering-Technology (BET) program on technology innovation at Auburn University. This book is the result of a course I offered in the BET program for about fifteen years while training hundreds of engineering as well as business students at Auburn University. I consider it a privilege to have served as the director of the Center from 2005 to 2014.

I also gratefully acknowledge the contributions of patent attorney, A. J. Gokcek, BSEE, JD, LLM, who served as a valuable resource for me over many years on matters concerning patents and the USPTO. He was my co-teacher for a few years in a course on Patent Application Drafting for Auburn University students. He served as an associate attorney at an IP law firm, director of IP at a major research university, and senior IP attorney at a US Department of Energy National Laboratory. He has personally drafted and litigated more than a hundred patents and patent-related legal cases, and, during his career, has counseled on all intellectual property matters including IP licenses, contracts, and litigation in the federal, academic, and corporate sectors. This book enables me to pass on to young inventors some of what I have gained from my association with him. Any errors in the book are, of course, mine alone.

CREATE VALUE IN THREE PHASES

1 | Engineers Create Value for Investors

When you ask an engineer what engineers do, the common response is that engineers solve and fix problems, design and model things and systems, create mathematical models, and many other variations along the same lines.

THINK: "I can Build Value for investors"

Activity (#)	$--Post-activity Value--average	
Good Idea (100)	1,000	not cashable
Product (50)	10,000	not cashable
Patent protection (15)	50,000	cashable
Business Model (7)	100,000	Workable?
Business Plan (5)	150,000	Cash flow?
Top management (2)	300,000	Right people?
Some Sales/potential sales (2)	1,000,000	Value/ investor

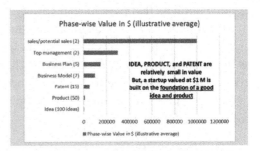

Figure 1.1. The million-dollar journey of an engineer's idea

Actually, engineers create value for investors as shown in Figure 1.1. To explain the table and bar graph in Figure 1.1, imagine that engineers produce 100 ideas: 50 of them may be reduced to individual products; 15 may be patentable or patented; 7 may support a WORKABLE business model; 5 may lead to solid business models and business plans with ADEQUATE five-year CASH FLOW into the future; 2 may have the RIGHT PEOPLE to manage the start-up (including the engineer-entrepreneur, most often); and finally 2 may have real or potential sales that could interest wealthy investors to value the business at a million dollars (Swamidass, 2014).

In this hypothetical journey, 100 ideas may produce two start-ups valued at a million dollars each by wealthy investors, who may invest in them as a result of their valuation. By creating value for investors, engineers could become entrepreneurs using their own ideas, while using investment capital from wealthy investors – a win-win situation for both the engineer and the investor; some engineers/scientists, not interested in starting and running a business, may license the technology to investors/licensees, and may let them start and run the business in return for royalties and/ or equity.

While engineers who work for an employer may think they are merely solving problems for their employers, they are actually creating value for their employers too; their employer benefits, and so do their customers.

Why This Book? What Is the Book's Theme?

Having taught hundreds of students from business and engineering colleges at Auburn University for 15 years using the contents of this book, it allowed the author to refine it into a volume that will adequately prepare engineers and engineering students to acquire the confidence necessary to recognize a commercially viable idea and graduate it to a tangible product, which in turn will lead to profit-making new businesses. Engineering students, with their college training, can create numerous products of value for investors and society at large.

Of course, not all engineers/scientists who invent must start their own businesses. There are plenty of investors who are eager to license promising products developed by engineers. Licensees with business and investment skills could start and run a new business, or add the licensed product under the umbrella of an already existing business that pays regular royalty to the inventor.

Each year, millions of engineers graduate from engineering colleges worldwide, and there are at least four times as many students in university engineering colleges as the number that graduate each year. Their creative potential is often lost to society because they lack the training and encouragement in their education necessary to develop ideas into inventions of not merely new products, but products that can create value for wealthy investors.

Business starts with an idea

- No idea --> no product
- No product --> no customer
- No customer --> no business
- "Customer is King/Queen"

Figure 1.2. Basic Principles of Business for Engineers.

Engineering students and engineers often scale back or even completely abandon their creative efforts because they mistakenly believe that they must have the investment capital to make their inventions become commercially successful. That is not true. If they have a strong idea, a commercially viable product, and patent protection, experienced investors can join them or take over the business part of delivering the invention to consumers. *This is the theme of this book.*

Figure 1.3. Thousands of customers in a long line outside the Apple store, New York City, September 2014, the day of the launch of iPhone 6.

(Photo © Cicchetti /Shutterstock. com, 218258131)

Engineers must pay close attention to the fundamental principles of business, presented in Figure 1.2. A product based on their idea must satisfy consumer needs, and thus attract customers (Figure 1.3), in order to create, in turn, a viable, profitable business. Without a product there can be no business. Therefore, engineers will be rewarded if they devote their time to creating products from their ideas that appeal to customers; if they do so, investors and their capital will follow.

Case Study 1

A Value Creator on the *Shark Tank* Show

Engineers unfamiliar with the idea of value creation for investors must watch the American TV series *Shark Tank*, which provides several good examples each week; archived shows can be viewed online at ABC and CNBC websites, YouTube, and other Web-based locations.

On *Shark Tank*, inventors receive investment to expand and grow their businesses, as well as support and advice from billionaire "sharks" *in exchange for equity* in the company started by the inventor. In return for a share of ownership in the start-up business given away to the "shark," the inventor receives access to their capital, expertise, network of manufacturers, retail outlets, access to US and international markets, and expert advice for rapid growth – all of them invaluable.

One of the best learning moments on the show occurred when Rick Hopper appeared on the show asking for $150,000 in return for 15 percent equity in his company for his very simple invention, ReadeREST, a device that secures eyeglasses or sunglasses to a shirt, coat, or other garment. The following figure shows his product, now sold on the Web as well as in several brick-and-mortar outlets. The video of the successful presentation on the *Shark Tank* show can be accessed on YouTube, ABC TV archives, CNBC archives or other archives for online videos.

In the archived YouTube presentation there are many valuable lessons for a professionally trained engineer as well as for anyone with even rudimentary engineering

Figure 1.4. ReaderREST (Amazon.com; January 2016).

skills. Whether Rick Hopper is an engineer with a college degree or not, his presentation shows that he has the skills to devise a solution using engineering principles. With that rudimentary product, he created value for wealthy investors such as Lori Greiner, one of the "sharks" in the *Shark Tank*, who invested $150,000 in Hopper's invention in return for 65 percent of his company; two years after that particular segment aired, sales of ReadeREST reached nearly $10 million, creating value for inventor Rick Hopper and investor Lori Greiner.

The facts of Rick Hopper's case:

1. A simple unsolved problem exists, namely, there is no good place to store glasses on one's person without using a case (bulky, uncomfortable) or unsightly devices such as "granny chains."
2. Hopper develops a simple solution using magnets and paperclips.
3. A patent search reveals to him that he is the second inventor with this idea; the full patent from the first inventor is reproduced in Chapter 13, Supplement 2.
4. There is no product in the market based on the original patent.
5. Hopper makes a great decision; using the published patent document he identifies the original inventor, contacts him, negotiates, and purchases the patent for $5,000.
6. Before Hopper appeared on the *Shark Tank* show, sales of ReadeREST had reached about $65,000, and he wanted to partner up with one or more investors from the show to grow and expand, nationally and internationally; he manufactures the items in his garage for about $1.05 each.
7. He sells them in a $9.99–19.99 price range on Amazon.com (Figure 1.4) and other Web-based retail outfits.

8. He asks for a $150,000 investment from the "sharks" for a 15 percent equity ownership stake in his company, ReaderREST, implying a company valuation at $1 million.

9. Lori Greiner, one of the "sharks" on the show, offers $150,000 for 65 percent of the company (valuation has been reduced by Lori Greiner to $231,000), with a promise to make him a millionaire soon.

10. Hopper has invested $5,000 to buy an issued patent and received $150,000 by giving up majority ownership in his own company and allowing Lori Greiner to control his company, but with a promise of much more income to come in the future.

11. He accepted the offer in 2012; by December 2014, sales of ReadeREST reached nearly $10 million; clearly Lori Greiner's capital and active participation have helped the company and its value grow.

12. Hopper has created value for Lori Greiner and himself through a simple invention that weighs less than two quarters (Figure 1.5).

Figure 1.5. ReadeREST next to two US quarters.

This case shows that commercial success is possible and within reach even with a simple idea and product. There is a lesson here for future engineer-entrepreneurs who may tend to make things too complicated for themselves: **You can build value with simple ideas and products, and you may let the investor turn your product into a large, viable, income-producing business.**

Case Study 2

GoldieBlox – a Massive Value Creator

Debbie Sterling, while a student at Stanford University, realized there were very few female students at engineering colleges or in engineering programs at universities (less than 11 percent when she was in college). She developed a passion for the idea of introducing girls to the joy of engineering at an early age. She could have chosen to do any number of things/products to put her idea into practice; she chose to create toys that are attractive to girls but with engineering content that is lacking in current toys for girls. Go to YouTube to watch her in a Kickstarter (Kickstarter is a crowd funding website that inventors use to seek funding from the public, without giving away ownership in the business/product) video promoting her new GoldieBlox line of toys aimed at preparing girls for a potential career in

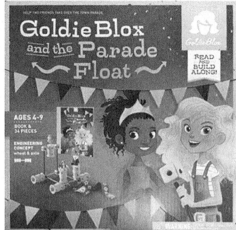

Figure 1.6. Examples of two boxed GoldiBlox toys.

(Photo © author)

engineering: www.youtube.com/watch?v=y-AtZfNU3zw (Note: the website may become obsolete with time)

From the video and Kickstarter website we note the following:

1. In converting the idea to a product, Sterling reasoned: "Girls like to read; boys like to build."
2. She designed an engineering-themed toy that helps a "girl inventor" build simple machines using a combination of spatial and verbal skills.
3. The toy was accompanied by a "girls-friendly" book and a tool kit.
4. The colors chosen for the toy were girls-friendly.
5. Sterling claimed she invested all her assets to make one working, high-quality prototype.
6. She tested it personally with more than 100 children and parents.
7. Sterling made an expensive video to promote her GoldieBlox toy on Kickstarter, seeking funds to place a manufacturing order for at least 5,000 units; she launched the project on September 18, 2012, seeking $150,000.
8. Four days later, on September 24, Sterling reached her funding goal; the toy went into production right away.
9. By October 18 – merely a month after launching her Kickstarter campaign – she had collected almost $285,000 from 5,519 backers, with an average contribution of $51.60.

Debbie Sterling's value creation with this product goes beyond purely monetary; she has enabled thousands of young girls to awaken their interest – and potentially their skills – in engineering via an experience acquired playing with a new line of toys. The societal value of her achievement is truly priceless. The author saw the Kickstarter video in 2013 and immediately ordered the original GoldieBlox toy for his granddaughter, who was seven years old at the time.

Of course, given the toy's target audience, we will have to wait about ten to fifteen years to find out if the young girls exposed to GoldieBlox are any more interested in engineering careers than girls not exposed to it. Regardless of the future results, however, Sterling has succeeded in starting with a difficult idea, turning it into a viable product, and eventually achieving great commercial success; the GoldieBlox line now sells widely both online and in brick-and-mortar retail outlets (Figure 1.6). Her products have been so successful that in 2014, Sterling's company could afford to run a 30-seconds advertising spot during Super Bowl – the most widely watched program in the United States practically every year, and thus carrying the highest price tag for ad time of any TV show (Said, 2014).

Compared to Hopper's idea presented earlier in the chapter, Sterling's idea to introduce girls to the joy of engineering at an early age is a complex and difficult idea to reduce to a product. But she was not swayed by decades of unsuccessful attempts of federal and state governments to attract female students to engineering programs; she chose to take on the challenge, and in the process created value for herself and her investors, as well as, of course, for girls and their parents. The growth rate and the sheer magnitude of her commercial success are very large because she took on a very difficult idea to commercialize. There is a lesson here for aspiring innovators: when a difficult idea is commercialized successfully, it brings great returns to the inventor; an extreme example would be an invention for generating inexpensive energy from sea waves. A commercially successful inexpensive wave-energy converter will be financially very rewarding to the inventor; while there are many inventions today for converting waves to energy, none hugely successful as of yet.

For quick success, take on a simpler product for commercialization to build up your financial capital to take on larger challenges later. Inventor-entrepreneur Elon Musk, founder of Tesla Motors and SpaceX, has projects that could be described as "going to the moon," literally, but he is backed by $14 billion personal wealth (as of 2016) that he has accumulated as a cofounder of PayPal.

TECHNOLOGICAL INNOVATION
AND INNOVATORS

2   Introduction to Technological Innovation

Innovation: What Is It?

The term "innovation" is used loosely today, and there is no widespread agreement on what it means. Renewal or change is implied by the Latin noun *innovatus*. While invention is something new, novel or unique, innovation is a new, unique or improved marketable product or service that can generate revenue from paying customers. A commercially successful innovation is a unique/new/novel product or service that can bring sustainable profits for an established business or a new start-up business.

(Trekandphoto/Adobe Stock, 75344522)

Some businesses use the term "innovation" to make their products appear competitive or "cool" without any real substance to back up their claim. Leslie Kwoh, in an article titled "You Call That Innovation?" says, "Just about every company says" it has innovation (Kwoh, 2012). Kwoh notes that 250 books with "innovation" in the title were published in a three-month period in 2015 based on a search of Amazon. com website. The article gives the following examples of innovation:

1. Creating a product that never existed.
2. Converting a waste or a discarded material to revenue-generating product.
3. Extending the scope and application of an existing product.

Innovation can be classified into (1) efficiency innovations that make products cheaper, faster and better; (2) sustaining innovations that improve existing products; and (3) disruptive innovations that bring new products to the market that have the potential to make existing products obsolete over time because of enhanced functions offered by the new product, decreased costs, availability, and other advantages (Christensen, 1997; Christensen and Raynor, 2003).

Technological innovation occurs when a technology-intensive product or service is developed and commercialized to produce sustainable revenue and income for the business. Apple Inc. has set an enviable standard for technological innovation for extant companies. Apple Inc. developed and introduced a commercially successful sequence of technology-intensive products such as the iPod, iPhone, and iPad into the market and achieved dominating status rapidly while generating exceedingly high revenues and profits for the company. When their new products were launched, they attracted long lines of customers waiting to get their hands on the new releases. In this book, "innovation" is addressed in the context of technological innovation by scientist/engineer-entrepreneurs, whose ideas and products create new business start-ups with sustainable revenue and profits.

Think of Engineers/Scientists as Value Creators

In this book, technological innovation is viewed as a journey in seven phases, stretching from an idea to a mature, innovative business. This phased view of technological innovation permits the highlighting of the unique and vital role of engineers/scientists during the first two phases of the journey; they may play an important role in the later phases too, but not as unique as their role in the first two phases.

Most graduating engineers/scientists have the skills needed to initiate and master the early phases of the technological innovation journey; because of a blind spot in their education, they are not aware of this. Furthermore, engineers/scientists need convincing that they are value creators in the technological innovation journey. Once they create value early in this journey, business professionals and investors can help them, as team members or independently, to turn the fruits of their efforts into a successful start-up business.

Unfortunately, most engineers/scientists rarely view themselves as value creators. They are unaware of the fact that they hold the key to new product/service ideas that can become the nucleus of the next small or big business. As students, most of them tend *not* to see beyond the acquisition of knowledge through lectures or in laboratory settings in numerous technical topics and subtopics such as dynamics, thermodynamics, hydrodynamics, and calculus, among numerous others.

It would be revolutionizing to the economy and to the personal wealth of engineers/scientists if more of them viewed themselves as value creators in the technological innovation journey. If they did, they would certainly give a boost to their job satisfaction, create numerous jobs and grow the economy.

Questions to ask yourself:

1. Do you want to invent?
2. Do you want to invent and innovate?
3. Do you want to create wealth for yourself and potential investors by inventing new products?

If you answered "yes," then this book is for you.

3 The Seven Phases of Technological Innovation

Why the Seven-Phased Model?

To introduce the idea of value creation by engineers/scientists, this chapter uses a seven-phased model of technological innovation in Table 3.1 as well as Figure 3.1 to communicate with engineers/scientists. To seasoned business professionals and investors, the two-dimensional, seven-phased model in Figure 3.1 and Table 3.1 may seem an oversimplification because of the model's artificial boundaries and strictly sequential progress over time. The intended audiences for this book, however, are engineers and scientists, who may not be familiar with even rudimentary business practices, start-up valuation, or investment. For them, I have found this to be an effective introduction to the unfamiliar.

The simplified, seven-phased view of technological innovation convey to the engineer/scientist the essence of the technological innovation journey from Phases 1 through 7, while making them understand and appreciate their unique role in the first two phases. An equally important message is: *engineers/scientists make a valuable contribution to the success of the later stages of a new business by the quality of their work in Phases 1 and 2 of a start-up business.*

Walter Isaacson, known for his book on Apple founder Steve Jobs, said something similar: "If you care about making good products, *eventually*, profits will follow."[1] In a new start-up, the quality of the efforts of engineers/scientists in Phases 1 and 2 ensures the "making of good products."

Convincing Engineers to Make a Start

One goal of the seven-phased view is to encourage hesitant engineers/scientists to kick-start the technological innovation journey with their ideas without being intimidated by the unknowns of Phases 3–7, which become relevant only when Phases 1 and 2 create new value in the form of new products/services. Phases 3–7 cannot exist without preceding Phases 1 and 2.

---

[1] Issie Lapowsky, "Walter Isaacson: 5 traits of the true geniuses," *INC Magazine*, http://www.inc.com/issie-lapowsky/walter-isaacson-5-traits-true-geniuses.html 3/3/2014 (emphasis added).

Table 3.1. *Seven Simplified Phases of Technological Innovation for Engineers and Scientists*

Phases 1–3: Pre-Start-up Phases	**Invention Phase 1: Conceive an Idea**
	An idea to solve a problem, meet customer needs or grab an opportunity; refine the idea
	Invention Phase 2: Develop a Product (or service) from the Idea – Value Creation
	Reduce the idea to a product or service for a target market, develop and improve it, protect the intellectual property
	Planning Phase 3: Business Model and Business Plan Development – Add to the Value
	Get help from mentors with entrepreneurship experience; form a management team, if possible. The business model, on paper, covers all aspects of the business to generate sales, revenue and profits. Business plan projects cash flow for three to five years assuming the business model is functioning; investors look for a strong product, business plan and cash flow for valuation purposes
Phases 4–7: Execution Phases – Financing the Start-up and Beyond	**Execution Phase 4: Find Investment Capital – Success Here Is a Function of the Value Created in Phases 2 and 3**
	Find capital (estimated time: 1–24 months); form a management team if not already done in Phase 3
	Execution Phase 5: Create a Functioning Start-Up
	Give legal and organizational shape to the start-up (estimated time: 3 to 30 months)
	Execution Phase 6: Grow to Financially Steady State or Maturity
	Growth in revenue and profits. Financially steady state/maturity (estimated time: 24–100 months); the engineer/scientist may sell ownership to exit the business now with a handsome financial reward.
	Execution Phase 7: Continue to Innovate Beyond Maturity
	Past maturity, yet remaining innovative; Apple Inc. is a good example. In 2014, the company was more than 30 years old and was yet considered very innovative in Phase 7.

Use Table 3.1 with Figure 3.1 below.

Another goal of the seven-phased view of technological innovation is to motivate the engineer/scientist to focus on Phases 1 and 2 without being paralyzed into inaction by the unknown challenges of Phases 3–7 that tend to nag them before they even start. Some of their fears are genuine but most are misplaced fears before their time – that is, before the completion of Phase 2. By never initiating Phases 1 and 2, engineers/scientists cause their ideas to fail even before they start.

Many engineers/scientists are not aware that potential investors are looking for value created by engineers/scientists, and would invest when they see value in new engineering/science products that have gone through rigorous Phases 1 and 2.

Technological Innovation: Not a One-Person Journey

Technological innovation is a journey. The person who starts with Phase 1 may not be around when Phase 7 is reached; therefore, this is not necessarily a one-person journey. The entire journey from a novel idea to a successful mature business may take years. The seven-phased view of the complete journey of technological innovation

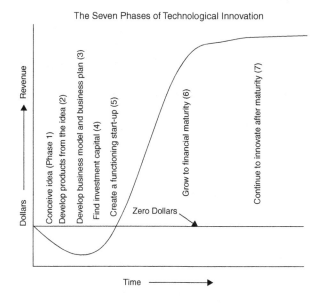

The Seven Phases of Technological Innovation

Figure 3.1. Invention phases 1–2; planning phase 3; execution phases 4–7 (use Figure 3.1 with Table 3.1).

(*Source*: Swamidass, P. (2014). "Engineers and Scientists: Value creators in the seven-phased model of technological innovation, *Technology and Innovation*, 16 (3–4).)

makes the journey easier to grasp and understand for new innovators who are engineers or scientists. The prioritization of the innovators' efforts concerning, what is immediate and what is not, is likely to remove the need for the aspiring engineer/scientist-innovator worrying prematurely about future phases of innovation before the early phases have been addressed adequately.

Pre-Start-Up Phases: Invention and Planning Phases

The seven phases fall into three distinct groups, as identified in Table 3.1 and Figure 3.1: **Pre-Start-Up Invention Phases** 1–2; **Pre-Start-up Planning Phase** 3; and **Execution Phases** 4–7.

During Invention Phases 1 and 2, the engineer/scientist plays a lead role or even a lonely role. Furthermore, during the invention phases, an idea is conceived and translated into a viable product/service for a specific target market. As the journey moves to Planning Phase 3 and beyond, the innovator may need the help of others with business and investment skillsets. A business model and ensuing business plan are parts of the finished product of the Planning Phase 3, which signifies the end of the Pre-Start-Up Phases (Table 3.1). If a viable marketable product or service does not emerge from Invention Phases 1 and 2, subsequent phases become irrelevant.

Caveat

A word of caution here: in real life, the lines between any two adjacent phases are blurred, and the activities of two adjacent phases may be iterative; for example, as the business plan makes progress in the Planning Phase 3, it may become necessary to go back to Phase 2 and improve the product by adding a new feature or by altering an existing function of the product. Therefore, while a sequential progress is implied in Table 3.1, technological innovation is never a one-directional progress from Phase 1 onward.

The reader is cautioned that the duration of each phase varies substantially from start-up to start-up. Consider the popular Amazon.com; it took more than six years for the start-up to make a profit (1994–2001), and a really small one at that. Some investors in the early days of the company were getting impatient. Here is an excerpt from Wikipedia:

> Amazon was incorporated in 1994, in the state of Washington. In July 1995, the company began service and sold its first book on Amazon.com .... In 1996, it was reincorporated in Delaware. Amazon issued its initial public offering of stock on May 15, 1997....
>
> Amazon's initial *business plan* was unusual; it did not expect to make a profit for four to five years. This "slow" growth caused stockholders to complain about the company not reaching profitability fast enough to justify investing in, or to even survive in the long-term. However, when the dot-com bubble burst at the start of the 21st century, destroying many e-companies in the process, Amazon survived, and grew on past the bubble burst to become a huge player in online sales. It finally turned its first profit in the fourth quarter of 2001: $5 million (i.e., 1¢ per share), on revenues of more than $1 billion. This profit margin, though extremely modest, proved to skeptics that Bezos' unconventional business model could succeed. In 1999, *Time* magazine named Bezos, the founder of Amazon.com, the Person of the Year.[2]

There are companies that become profitable much sooner; most successful start-ups involving simple products/services bring in profits in one to three years. More complex products take longer to be profitable; Amazon.com is an example of a complex new business because there never was an Internet-based retail company on the scale the founder of Amazon.com contemplated.

Cash-Flow-Based Graphical Representation of the Seven Phases

Figure 3.1 is a two-dimensional plot of **cash flow** during the seven phases of technological innovation, from an idea to a stable, mature business. In this figure, cash flow ties all seven phases into a common journey, making cash flow the common thread. The figure shows approximately where the individual phases occur over time as well as a **curve depicting cash flow in dollars from the start of the idea**; the cash flow in Phase 7 could be much higher than shown in the figure if the company remains technologically innovative in a manner similar to Apple Inc. A company's market valuation (i.e., share price times the shares in the market) is tied to current and potential cash flow in the future; in 2012, Apple Inc. became the most valued company on the US stock market on the strength of its technological innovation capability, past, present, and future.

Negative Cash Flow during Pre-Start-Up

According to the cash flow curve in Figure 3.1, a business produces positive cash flow after a functioning start-up (Phase 5) begins to make sales. Until sales bring revenue,

---

[2] Accessed June 13, 2014 (emphasis added).

the cash flow is negative during the pre-start-up activities; *cash infusion from investors during pre-start-up is not shown in the figure.* During pre-start-up, investment to cover the negative cash flow could be the inventor's own funds, or funds from family and friends.

The pre-start-up negative cash flow could be very small for some businesses; certain software and Internet businesses are examples. Yahoo! search engine created by two college students in the 1990s was a grand success on a small personal budget – Yahoo! became an "overnight" success, and venture capitalists came knocking on its door because of its cash flow potential.

Start-up businesses fail frequently. In very simple terms, failure of a start-up company is said to have occurred when the company fails to make positive cash flow; that is, revenue from sales does not cover all the costs and start-up expenses. That could be due to a number of reasons, the foremost being not enough customers to purchase the product or service at a price that would produce a positive cash flow.

The rate of failure among start-up companies is reported to range from 30 percent to 95 percent depending on the metric used. Debra Gage quotes a study by Shikhar Ghosh of Harvard University: "If failure means liquidating all assets, with investors losing all their money, an estimated 30% to 40% of high potential US start-ups fail.... If failure is defined as failing to see the projected return on investment ... then more than 95% of start-ups fail." The article notes, "Of all companies, about 60% of start-ups survive to age three and roughly 35% survive to age 10."[3]

The Importance of Pre-Start-Up Phases 1–3

Weak or shoddy work by scientists/engineers in the first two phases could prove fatal to a start-up business. Business professionals who may contribute to the success of the start-up after Invention Phases 1 or 2 sometimes may be unable to correct the handicap inflicted by an ill-developed product that emerges from the invention phases. Even if the engineer/scientist overcomes the inadequacies of the invention phases by changing and improving the product during the later phases, valuable time and investors' money may be lost in the process.

The science/technology innovator with a new idea must ensure that he/she does the "homework" during Invention Phases 1 and 2 to increase the success of the start-up business. For some products, after moving beyond Invention Phases 1 and 2, the "die is cast," and it may become impossible or too expensive to change the product, its features, or its appeal; this may partly explain why some businesses fail. This is why successful engineer/scientist innovators with a new idea for a start-up are "obsessed" with customer-focused product development during the Invention Phases 1 and 2 to lay a sound foundation for subsequent phases.

---

[3] Debra Gage, "The venture capital secret: 3 out of 4 start-ups fail," *The Wall Street Journal,* September 19, 2012.

4 Engineers Add Value in Stages

Value Creation by the Engineer/Scientist-Entrepreneur

Once an engineer/scientist arrives at the end of Invention Phase 2, he or she sub-stantially increases the chances of success for a business based on the outcome of a rigorous Phase 2. At this stage, the engineer/scientist enhances the dollar value of the original idea many times over. For example, if the original idea at conception was worth $2,000, the value of this idea once it is refined and reduced to a working proto-type, protected by a Provisional patent application to the USPTO, could have grown to $50,000 or $100,000 (pre-investment valuation) for potential investors looking for a technological innovation for investment. Investors use pre-investment valuation of the start-up business for negotiated partial ownership in return for investment in the start-up business. Granting a partial ownership of a company to an investor is a legal process; it involves a legally created business entity with the ability to give ownership in the company in exchange for investment in the company by others.

How an Engineer's Idea Grows in Value in Early Phases

If the pre-investment valuation of the start-up is, say $250,000 after Phase 2, it means that 20 percent ownership in the company would require a $50,000 investment from investors. Thus, the value created by the engineer/scientist at the end of Phase 2 can be converted to investment in the company in exchange for partial ownership in the company. It is not uncommon for an investor to buy out the entire start-up by paying the full valuation amount. After Invention Phase 2, the value of the start-up reflects several value-enhancing items such as:

a) the value of the original idea;
b) the idea's value amplified by the product developed from the idea;
c) significant target market;
d) a working prototype;
e) granted patents or patent applications to protect the intellectual property (IP).

This progressive value enhancement of an idea occurs in Invention Phases 1 and 2 without the assistance of outsiders such as business professionals or investors. However, the addition of qualified, experienced business professionals to the management team after Phase 2 could further enhance the value of the start-up. Because of the background and experience business professionals may bring to a start-up, they may enhance the chances of success of the business, thereby causing investors to assign a higher pre-investment value to the start-up. If we continue with the earlier example of a start-up valued at $250,000 at the end of Phase 2, with the addition of experienced business professionals, the start-up may be valued at $500,000; an investor must then invest $100,000 in return for 20 percent ownership in the start-up.

With the addition of strong business professionals, a validated business plan could emerge during Planning Phase 3. A validated, strong and convincing business plan could add to the valuation of the start-up business. If we continue with the earlier example, the pre-investment valuation may reach $750,000 once a strong and valid business plan is in place showing estimated cash flows for three to five years in the future. If so, an investor must invest $150,000 in return for 20 percent ownership in the start-up company.

The message of the preceding paragraphs to an engineer/scientist-entrepreneur is: do not hesitate to develop a good idea, reduce it to a practical and viable product, apply for patent (at least a Provisional Patent application), and develop a prototype during Phases 1 and 2. Never mind Phases 3–7 when Phases 1 and 2 have not been completed.

Remember, only an engineer/scientist can multiply the value of an original idea during Invention Phases 1 and 2; business professionals and investors can only add to the value created by engineers/scientists.

5   Disruptive Technological Innovators: Value Creators

Disruptive technological innovations disrupt established businesses; digital camera is an example, as it ended Kodak's preeminence in chemicals-based photography.[1] Such innovations are rare. Disruptive innovations such as personal computers first introduced in the 1970s, Internet-based businesses introduced in the 1990s, and others have created trillions of dollars in new markets over the past few decades.

Disruptive innovations may be new products/services that either slowly create vast new markets or replace established products, or both. Clayton Christensen introduced the idea of disruptive innovations.[2] The idea is now well recognized, and some companies actively seek disruptive innovations. Such innovations bring the "next great wave" of changes to an industry and have the potential to wash away established businesses.

Kodak, the king of old-style photography, was toppled by the digital substitute despite Kodak's efforts to stay ahead while ignoring the slowly emerging disruptive digital technology. Disruptive innovations are getting a lot of well-deserved attention among growth-conscious businesses for two reasons:

1. Corporations want to capture and rule vast new markets to ensure systemic corporate growth; Apple Inc. demonstrated how this could be done by entering the iPod (music player), iPhone (smartphone), and other market niches new to the original, computer-oriented company.
2. Corporations want to protect against potential fatal threats to their current markets that are vulnerable to the inevitable next wave of disruptive innovations.

The individuals behind disruptive innovations are often fascinating characters: Thomas Edison, Steve Jobs (Apple), Bill Gates (Microsoft), Jeff Bezos (Amazon),

---

[1] A version of this chapter originally appeared in American Management Association's *AMA Shift*, Leader Board Blog, "Hunting for disruptive innovations? Try looking for innovators," by Paul Swamidass, March 23, 2012, http://www.amanet.org/shift/index.php/2012/03/23/hunting-for-disruptive-innovations-try-looking-for-innovators/ (accessed July 14, 2014).

[2] Clayton M. Christensen, *The Innovator's Dilemma*. Boston, MA: Harvard Business School Press, 1997.

Sky Dayton (Earthlink), Marc Andreessen (Netscape), Jerry Yang/David Filo (Yahoo), Larry Page/Sergey Brin (Google), and Mark Zuckerberg (Facebook) have all been successful, but also controversial. These "disruptive" innovators are credited with technological innovations that disrupted the market for established businesses and/or created totally new unforeseen massive markets.

The Importance of the Innovator

Disruptive ideas began with individuals who, by chasing a personal vision with generous helpings of determination, perseverance, tenacity and focus, translated disruptive ideas into large-scale commercial successes. Individual innovators are often the engines of disruptive innovations. They have brought to market bold new ideas or products/services that others considered laughable, impossible, or noncommercial.

During a recent lecture, I asked my students: "Before iPhones became popular, if someone, who worked at the research labs of one of the conventional cell phone manufacturers such as Nokia, had proposed a new phone idea identical to the original iPhone, what might have been his or her chance of getting the idea approved?" Most commented that the iPhone, compared to the norm, was heavy, too big, too expensive, too different, brick-like and worked differently. Therefore, they concluded, the idea would have been rejected outright at the lab level. Yet Apple's iPhone made history; the idea needed a *disruptive champion innovator* such as Steve Jobs.

In 1977, computers were very large, and were exclusively used by large- or medium-sized businesses and government agencies. When someone told Ken Olsen, the founder and CEO of DEC (then a minicomputer leader), that a couple of guys were introducing a small microcomputer that could be used by individuals in homes, he was quoted as saying, "There is no reason for any individual to have a computer at home." Too often visionary ideas get laughed away or derided.

While businesses have started hunting for disruptive *innovations,* they have not focused their attention as much on disruptive *innovators*. Without disruptive thinker-innovators there will be fewer disruptive innovations.

Tests for Disruptive Innovators

Disruptive innovators make fascinating reading. The 2011 book on Steve Jobs by Walter Isaacson is an example.[3] Disruptive innovators:

1. Withstand amused rejection of their pet ideas for a long time (let's call it the "laugh test").
2. Handle 99.9 percent opposition or worse.
3. Are paranoid about threats to their work and markets (like Andrew Grove of Intel[4]).

---

[3] Walter Isaacson, *Steve Jobs*. New York: Simon and Schuster, 2011.
[4] Andrew Grove, *Only the Paranoid Survive: How to Exploit the Crisis Point that Challenge Every Company*. New York: Doubleday, 1996.

4. Know that if they can't make their idea work, no one can.
5. Know that if they work for an employer, the employer may have no interest in their idea. This is why Steve Jobs recommended owning a company in order to have total control over the future of your ideas and products. When he was fired from Apple, he just founded more companies.
6. Know that if the success of their idea depended on others entirely, it might soon die (Sir James Dyson, inventor of a now world-renowned vacuum cleaner, made over 5,000 prototypes, while others might have quit earlier – he did not quit[5]).
7. Know that making their idea work should be their goal and responsibility, and no one else may care or know enough to make it a success.

Most innovators are not disruptive innovators, but they can learn from a disruptive innovator. Standout technological innovators such as Steve Jobs, Sir James Dyson and Jeff Bezos excel in technical as well as business matters.

Businesses, if they are serious about embracing disruptive technologies, must not only identify disruptive innovators but must also develop a corporate culture to accommodate them and let their ideas evolve to their natural conclusion; this is a daunting task. These individuals aren't going to fit in easily with traditional corporate rules and strategies. Most corporations would likely screen out such persons at the interview phase: it is no surprise that many of these individuals set up their own companies.

If companies are hunting for disruptive innovations, they must learn to embrace disruptive innovators: you can't have one without the other.

---

[5] Sir James Dyson, *Against the Odds: An Autobiography*. New York: Texere LLC, 2003; Margaret Heffernan, "James Dyson on creating a vacuum that actually, well, sucks," *Reader's Digest*, February 2009, http://www.rd.com/advice/work-career/james-dyson-on-creating-a-vacuum-that-actually-well-sucks/.

PHASE 1
CREATE VALUE THROUGH A NEW IDEA

FINDING YOUR IDEA

6   Ideas: How Do You Find Them?

An Idea Gets the Ball Rolling

For Rick Hopper and Debbie Sterling, the two inventors discussed in an earlier chapter, an idea has gotten the ball rolling. When an engineer grabs an idea and gives it shape, the idea could become a tangible commercial product. Only then can value-creation begin for the engineer-innovator as well as for future investors.

If you are naturally creative and generate many new ideas frequently, this book should help you capitalize on those ideas. If you want to generate many new creative ideas, you can use proven methods described here.

(© Dizain / Shutterstock, 282852974)

Ideas: An Engineer's Strength

A creative engineer/scientist may come up with new ideas for products and services when he/she sees a problem needing a solution, or an opportunity for a new product in real life. Next, they need to function as "inventors." Inventors display the following qualities that Walter Isaacson, a prominent biographer of Steve Jobs, saw in his famous subject[1]:

1. A controlled passion for perfection.
2. A keen eye for details.

---

[1]   Walter Isaacson, *Steve Jobs*. New York: Simon & Schuster, 2011.

3. The love of simplicity.
4. Thinking different.
5. Thinking out of the box.

Where Do Ideas Come From?

Ideas come naturally to some, but others can turn to well-known sources for ideas:

1. **Problems begging for a solution:** example – lost keys, wallets, and the like. The solution is the key finder that sounds an alarm through your cell phone when you are separated from your keys. In addition to keys, it can be attached to anything else, such as a wallet or luggage (Figure 6.1).

Figure 6.1. SwiftyFinder BF-127

(Amazon.com, 2016)

Another example of problem begging for solution: leaves clogging the rain gutter (Figure 6.2a). It calls for an invention of an easy-to-install downspout wedge that prevents clogging (Figure 6.2b).

Figure 6.2a. Roof rain gutter

(© Suzanne Tucker / Shutter Stock 114347572)

Figure 6.2b. Downspout wedge
(Amazon.com)

2. **New opportunities for products:** the arrival of the smartphone offers thousands of new opportunities for products including, smartphone games such as the popular Angry Birds game (Figure 6.3).

Figure 6.3. Photo (Photo ©
author, February 2016)

3. **Substitute for old products in the market (Figure 6.4a):** the jar-shaped humane mousetrap. Mouse trap is set with food inside the bottom and neck is tilted to touch the floor. It tilts to a vertical position by the weight of the mouse that enters the trap – mouse gets trapped alive inside the long-necked jar in the vertical position without harming the mouse which can be released far away (Figure 6.4b).

Figure 6.4a. Old mouse trap

(© Canstockphoto / Guarding 7941571)

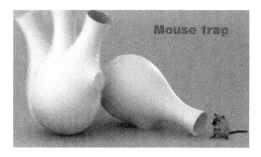

Figure 6.4b. The humane mousetrap. International design excellence award, IDEA 2010, student design. Awarded by the Industrial Designers Society of America (IDSA)

(©Philipe Vahe, used with permission)

4. **Connecting random things to find a new creative idea**: at first glance, it doesn't look like there can be anything in common between the words "watermelon" and "cube." But growing watermelons in a transparent cube results in a cube-shaped watermelon; easier to ship and allows more to be shipped in a single container, compared to regularly shaped watermelons (Figure 6.5).

Figure 6.5. Creatively grown watermelon

(© 1365Art: Order 1072351659)

5. **Deliberately thinking different:** the old-fashioned way of picking up tennis balls is tiring and time consuming during practice (Figure 6.6); it is much easier and more efficient to use a specially designed robot (Figure 6.7).

Figure 6a. Picking up tennis balls is tedious

(© Luril Sokolov / Adobe Stock, 51358510)

Figure 6b. Who can pick up these balls for me?

(© Carlos Horta / Shutter Stock, 125668034)

Figure 6.6. Picking up tennis balls is tedious

Figure 6.7. A prototype of the Tennibot, a robot that picks up tennis balls from the court while you relax

(Courtesy of © Tennibot LLC March 2016; used with permission http://www.tennibot.com/tennibot)

Deliberate Observation to Stimulate Idea Generation

It is harder to come up with new ideas in a vacuum. Priming the mind with idea stimulators can be accomplished easily; seek out and observe thousands of problems that have been solved by existing products currently on the shelves at large retailers, both Internet-based and brick-and-mortar ones.

1. Visit Walmart, Home Depot (Figure 6.8) and other stores and observe their shelves for ideas.

Figure 6.8. Inside a Home Depot store

(Photo © author)

2. Browse the offerings from various Internet retailers.
3. Browse patents and patent applications; this can be accomplished by using Google Patents or going to the USPTO website.

When you observe extensively with the intention of finding ideas, be a good critical observer: do not be limited by your assumptions; do not let your old assumptions kill creative new ideas (cube-shaped watermelon); and do not limit your options for problem solutions – "the more the merrier."

Do Not Discard "Simple" Ideas

In an earlier chapter you saw how Rick Hopper's simple idea created much value for him as well as a *Shark Tank* investor. Read "10 simple product ideas that made billions" by GrowAmerica to grasp the power of simple ideas that became products with significant market demand because consumers valued these simple products.[2] Examples featured by GrowAmerica include:

---

² GrowAmerica, "10 simple product ideas that made billions," http://learn.growam.com/10-simple-product-ideas/ (accessed July 2013).

1. Post-It notes, invented by Art Fry in 1974; generated $1 billion in revenue in 100 countries in 2012 (3M company product).
2. Band-Aid, invented in 1921 by Earle Dickson; generated $3,000 in sales in its first year; by 1961, Johnson and Johnson sold $30 million a year, and sales keep growing.
3. Slinky, invented in 1943 by Richard Jones; first introduced in 1945 just in time for Christmas. By 2012, James Industries has produced and sold more than $250 million in Slinkies worldwide.
4. Frisbee, invented by Walter Frederick Morrison in 1937; sold to Wham-O in 1955; has sold more than 300 million by 2012.
5. Velcro, invented by George Mestral in 1941; today it is a multimillion-dollar business, with Velcro sold in 40 countries.

Invention Phase 1 tests the creative side of the inventor. If the reader wonders what it means to be creative, the reader may challenge himself/herself to answer questions in the article, "15 most ridiculous college application questions."[3] Here is a sample of questions (there are no right answers):

1. What are unimagined uses for mustard? (University of Chicago)
2. Using a piece of wire, a car window sticker, an egg carton, and any inexpensive hardware store item, create something that would solve a problem. (Johns Hopkins University)
3. What invention would the world be better off without? (Kalamazoo College)
4. If you are reduced to live in a two-dimensional world, what would be your greatest problem? Greatest opportunities? (Hamilton College)

Brainstorming for Creative Thinking

Brainstorming is an effective tool for creative outcomes during Invention Phase 1. Brainstorming assumes there could be numerous potential solutions to a problem, while the goal of brainstorming is to find one "best" solution among numerous options. Ralph Keeny offers help in the search for the "best" solution.[4]

Do not limit your solution alternatives during creative thinking. Often brainstorming comes down to selecting among alternatives. Keeny distinguishes between *constrained* thinking and *constraint-free* thinking. An example of *constrained* thinking he mentions is the national debate in the 1980s in the United States to reset the freeway speed limit of 55 mph after the oil crisis ended. The debate focused on speed limit options of 55 mph and 65 mph only, which was further complicated by the desire to reduce annual fatalities caused by vehicle accidents; at 55 mph, there were 10,000 fewer recorded deaths each year. Keeny calls this an example of an

---

[3] "15 most ridiculous college application questions," http://finance.yahoo.com/news/-the-15-most-ridiculous-college-application-questions-160531465.html?page=all (accessed July 2013).

[4] Ralph L. Keeney, *Value-Focused Thinking: A Path to Creative Decision Making*. Cambridge, MA: Harvard University Press, 1996.

"incredibly limited range of alternatives." He offers additional alternatives that would have allowed for the reduction in the number of fatalities while going with the higher speed limit, such as:

1. Additional driving requirements for teenage drivers.
2. Additional requirements for seniors with declining vision or skills.
3. Regulation and enforcement of the use of child safety seats.
4. Requiring helmets for children.
5. Harsher enforcement of drunk driving.
6. Harsher enforcement of speeding drivers.

In the above example, if reducing annual driving fatalities is a goal, Keeny's message is, do not limit your decision alternatives to just regulating vehicle speed. *The danger is: you may leave out many better alternatives or the best alternative.*

Require individuals to generate alternatives and solution(s) before taking them to a group. Keeny is a proponent of this idea because he fears one person in the group or one idea (an "anchor" idea) in the group may dominate the discussions while preventing good ideas from others from emerging. Therefore, it is better for individuals to come up with their own list of solutions individually, and then to take the list to a group brainstorming session. This will ensure that more good ideas will be on the table for the group to discuss.

PHASE 2
CREATING A PRODUCT

7 Turn Your Idea into a Product

Viable Products Spring from Only a Few Ideas

Many ideas do not support a viable product; discard them and search for ideas that hold a viable product with a future. More than half of the ideas may be discarded quickly; about 15 percent of your ideas may lead you closer to a viable product, and perhaps only 5 percent may get you close to a product with a potential commercial value – this is what you are looking for. Learn to comb and search your own ideas carefully for gems loaded with potential value. Not all ideas give birth to products.

(© Dizain / Shutterstock, 246789448)

Example of an Award-Winning Product

In the 2015 Tiger-Cage contest at the Auburn University's Raymond J. Harbert College of Business, Thomas Walter Center for Technology Management of Auburn University recognized the most innovative products in this contest for student teams. The $1,500 award-winning product was the Tennibot, a

Figure 7.1. Tennibot prototype in the top picture is searching and picking up balls. The two lower pictures show the same prototype in a tilted position for feeding balls to the player

(www.tennibot.com/; © Tennibot, LLC; used with permission)

tennis-ball-collecting robot, invented by Haitham Eletrabi, and developed with a team of students at Auburn University.

Haitham Eletrabi is a tennis player, and he has witnessed both coaches and students pick up hundreds of tennis balls scattered all over the court every day. He thought that valuable coaching time was being lost on picking up balls instead of working with students. This was bound to improve if coaches didn't have to spend an inordinate amount of time at every training session picking up loose tennis balls.

Tennibot is the product born of that idea. It is a robot that finds and collects tennis balls, gathering them into a ball hopper for immediate reuse by a player or coach (Figure 7.1).

The inventor and his team exhibited Tennibot at the CES-2016, the world-famous Consumer Electronics Show held in Las Vegas, perhaps the largest in the world; it attracted attention from players, coaches, sports equipment retailers, and investors from the United States, Europe, and South America.

The Most Innovative Product Award given to Tennibot had the following criteria for selection. This list should help anyone in the process of developing a winning new product from an idea. The product:

1. has an element of surprise;
2. solves a challenging problem;
3. is an example of simplicity;
4. is creative and novel;
5. provides magical customer experience;
6. has an appealing design.

Products Are Defined by Multidimensional Functions

To develop a product from your idea, ensure that the new product has functions and features that consumers want and would pay for. Without at least one strong function, the product may have no commercial value. Functions have many dimensions. To be a well-desired product, it must excel in many dimensions, such as:

1. The problem(s) it solves
2. Performance
3. Efficiency
4. Cost and affordability
5. Produceability
6. Quality
7. Safety
8. Durability
9. Customer appeal
10. Novelty

A good product will satisfy most of the items in this list. A product in the early stages of development could be strengthened by the inventor by getting inputs from:

1. potential customers' survey, focus groups, etc.;
2. experts on the product;
3. Internet;
4. firsthand observation of competing products in use;
5. brainstorming to find new and better solutions.

Products Have Features too

Features do not alter the functions of a product. For example, decades ago, cars did not come with cup holders inside the car in various locations for the convenience of the driver and passengers. Today, all cars come with built-in cup holders. They do not alter the primary function of the car, but they are a desirable feature that every car buyer wants. The following stroller comparison is a good example of the same product with basic vs. enhanced features (Figure 7.2).

Basic version of stroller
Kmart, Model # US087BJV (April 2016)
Cosco Cupcake Umbrella Baby Stroller

Featured version of stroller
Kmart, Model # CV251BSY (April 2016)
Cosco Toddler Umbria Stroller - Twister

Figure 7.2. Featured version has a hood, child snack tray with cup holder, parent tray under seat, etc.

Use the Customer as Your Design Partner

As an engineer, once you have identified a product for full and complete development, do not let your engineering skills and sensibilities *alone* dictate product design and development; the better engineer you think you are, the more likely you are to make this common mistake of depending solely on yourself to design and develop a product. If you do so, you may end up with an excellently engineered product, but it may not be the one that customers would pay for, or not enough customers would pay for; in either case, you cannot build a business around that product. Therefore, if you can get the following inputs from potential customers, ask them:

1. Can you define the problem you want solved?
2. How are you currently solving the problem?
3. What are the strengths and weaknesses of your current solution?
4. How would you prefer to solve the problem?
5. What would you pay for a product that solves your problem?
6. What functions and features do you want?
7. What would be the cost for you if the problem is left unsolved?

Practical New Product Development

During Invention Phase 2, the idea from Phase 1 is reduced to a tangible and viable product that has customer appeal and is targeted to a specific market. A commercially viable product is one that can be produced, packaged, distributed, and enough of it sold through various marketing channels at a price that would result in healthy sales revenue and a reasonable profit.

The following activities constitute product development for most manufactured products during Phase 2. Some of the activities on the list undergo multiple iterations before finalizing the product during Phase 2. However, later phases such as Phases 3–6 may impose additional changes on the product due to market realities that become evident as the business takes shape.

When the business becomes operational, the product is constantly refined to adapt to changes to corporate strategy, market realities, competition, and customer inputs. Here are some of the major activities that constitute product development in Invention Phase 2 (a manufactured product is assumed – some items in the list would be different for software or service innovations):

- **Spell out the goals for the product:** What does it do? How does it function? What problem does it solve?
- **Develop the functions of the product**: function refers to what it does.
- **Develop the features**: for example, for a new cell phone such as the iPhone, the list of features could include the life of the internal battery, camera, colors, and so on.
- **Ascertain target customers and their needs**: without a customer, the best engineering/scientific product has no future in the innovation journey. You need to know your potential customer. Therefore, speak to a number of potential customers, or conduct a survey of potential customers to get customer input (this is called *primary data* that you gather from potential customers), to identify the functions and features desired by the potential customer; a young new innovator who is an engineer/scientist may consult an experienced business professional, who could serve as a **mentor** in gathering customer input, or for designing, conducting and interpreting customer surveys.
- **Design and develop specifications for components and performance**: for example, for the iPhone, components and parts such as the case, battery, key pad, antenna, software operating system, and so on must be developed and specifications selected (example: battery specification would include battery dimensions, voltage, power, life of battery, time to charge, time to discharge, etc.).
- **Develop drawings, dimensions, specifications, and tolerances:** for each component and assembly.
- **Select materials for parts:** some materials are chosen for strength, some for appearance, some for electric conductivity, and so on.
- **Select assembly and parts manufacturing processes:** this could be approximate in the early stages of development, to be refined as the business takes shape.
- **Find suppliers*:** for parts that are likely to be purchased or for acquiring a fully assembled product from suppliers, who can make quality products at a competitive price and deliver them reliably to customers on time; for example, when the iPhone was launched, the decisions would have included who would make and supply batteries and who would make the unique touchscreen glass.

- **Apply for patents to protect the intellectual property:** you may know that the iPhone has many patents and Apple Inc. has frequently gone to court to protect the iPhone against patent infringement by other companies that may want to use their components and ideas without a legal agreement with Apple; the best-known infringement cases are between Apple Inc. and Samsung.

*Suppliers can make or break a new product. Corning Inc. had developed a chemical in the 1960s to produce Gorilla glass (a proprietary glass). Without a market for the glass, the product was shelved and production discontinued. More than forty years later, Steve Jobs, head of Apple Inc., while looking for a touch-sensitive glass interface for the iPhone, during product development, noticed the suitability of Gorilla glass for the new iPhone still on the drawing board; he persuaded the head of Corning to restart production of Gorilla glass for iPhone decades after the product was shelved by Corning.[1] Corning, as a supplier, was key to translating one of the vital features of the iPhone idea to a practical product.

During Planning Phase 3 (business model and business plan development), some of the above steps of Phase 2 may need revision or reiteration. Furthermore, some steps of Invention Phase 2 may be reiterated in the execution phases, when the product is prepared for manufacturing.

Prototype construction and testing: Depending on the product and the nature of the business, a prototype may be produced and tested for performance as early as Phase 1, if the cost of producing and testing is affordable. If the investment needed to build and test prototypes is significant, prototype building may be delayed until Phases 4 or 5 when investment comes in during the execution phases.

(© Dizain / Shutterstock, 246641980)

The Importance of Customer Input before You Design

This is what Jeff Bezos, founder of Amazon.com, said in 1998 when asked to list the keys to success: thoroughly evaluate the market (customers make up the market), figure out what your customers want and then deliver it better than others. This is

---

[1] Walter Isaacson, *Steve Jobs. New York:* Simon & Schuster, 2011.

the only way to succeed, but it is also very hard to execute; Amazon customers want selection, ease of use, convenience and good pricing – "so we are making sure we offer them the best of all those."[2]

An engineer or scientist may find the idea of including customer input in product design rather unusual. They may mistakenly think that a brilliantly designed product by the best engineers or scientists must appeal to all customers. Unfortunately, that is not always the case. It is better to find out what the customer would want before devoting time and resources to designing, making and selling a brilliantly engineered product that very few, if any, would want.

Engineers and scientists entertaining the thought of making new products for commercializing their ideas must get used to the idea that "customer is king." Without customers there will be no sustainable business, regardless of the quality and excellence in engineering and science involved.

A product of brilliant engineering and/or science may deserve admiration but may not attract paying customers; if so, it would be considered a failed product and the cause of any related business failure. The best way to increase the probability of customers buying a product is to get **formal and reliable customer input** before product design and development are completed.

The price that customers would be willing to pay for the product is vital information to have before the design is completed. The features customers would want in the product would be important information also. These are just two of the many pieces of information a designer could gather from potential customers.

The Survey

How do you find out about potential customers' needs and tastes? There are many options for doing it. However, the most common method is to speak to a large number of potential or target customers and/or gather their input via a well-designed survey questionnaire. One of the many advantages to using such a questionnaire to gather potential customer input is that it ensures that all respondents hear (or read) the same questions. If all respondents are not hearing or reading identically worded questions, their responses could vary, and consequently mislead the designer.

The findings of a survey may be surprising to an inventor. Even after the product design is finalized, the designer should continue to seek information on customer tastes, likes and dislikes.

Getting Reliable Customer Input

It would be a mistake to use customer inputs from too few respondents. A reliable customer input means that the findings from a survey would be the same (or comparable) if a similar group of customers were surveyed later. This kind of *reliability* of the results of the survey is necessary before the designer places his/her trust in survey results.

---

[2] *Success* Magazine, July 1998.

Reliability of customer survey results is a function of the quality of the questions (example: unambiguous wording), representativeness of the potential customers who completed the survey, and the number of potential customers surveyed.

Number of Customers Surveyed

Reliability improves with the number of representative customers surveyed; with all other factors remaining the same, more respondents to a survey means a more reliable survey and its findings. A survey of less than 30 responses is not regarded as reliable unless the respondents are part of an expert panel with in-depth knowledge of the market, customers and products. A sample of about 100 representative potential customers is desirable.

Other Inputs to Seek

While customer inputs are powerful, the voice of experts may be useful too. Get input from experts who can speak on the product, customers, market and so on. A few recognized experts from the industry may be consulted, either personally or through their publications.

8 Early Detection of Market Potential

A. David Mixson

The Current State of Market Research for Entrepreneurs

The term "market research" has different connotations for entrepreneurs. In the late 1990s and early 2000s, some advocates recommended an early, quick and dirty assessment of the potential market to help entrepreneurs avoid wasted time and effort. Steve Blank together with Bob Dorf (Blank and Dorf, 2012) and Erick Ries (2011) are recognized for their customer development methodology to test an idea's validity and to determine the ideal target customer during the very early stages of a product or business idea's development. Ultimately you do not have a business if the idea is not valid and if there is no viable target market. Both Blank's and Ries's methodologies involve "getting out of the building" and talking to dozens of prospective customers using quick, cheap cycles of data collection and learning. The Eureka Ranch's (2015) Innovation Engineering (IE) process and the Business Model Canvas of Osterwalder and Pigneur (2010) are also contemporary techniques for early market-potential detection.

Reducing Your Market Risk

As an inventor with an idea or a product concept, the first task at hand is to ensure that you are addressing an important need, problem, or opportunity. Market risk arises when potential customers do not care about the problems your idea/product solves and the benefits it provides. Market risk also arises during the process of getting the product to your target audience.

Inventors need to address these risks early in the idea stage, even before the product is fully developed. This can be done by conducting early detection of market potential using quick cycles of data collection and learning.

David Mixson is a Certified Innovation Engineering Blackbelt and Growth Coach helping communities and their new business start-ups using lean start-up methodologies; contact the Government and Economic Development Institute (GEDI) at Auburn University.

One of the approaches used and recommended by the Eureka Ranch's IE process helps the inventor assess the size and frequency of the problem that the inventor is attempting to solve. The process works like this:

1. Identify several specific target audiences that could benefit from your idea; examples of target audience are: college students, home owners, males, etc.
2. When you have identified possible target audiences, both IE and the Lean Start-up methodologies recommend that you form a hypothesis (educated guess) concerning the potential or actual problems of the audience your invention is intended to solve – examples follow.
3. Subject the hypothesis to a validation test using data collected from the target audience.

Qualitative research: Interview people you think your idea may help (i.e., target audience) and get some initial insight into the problems they encounter. For instance, if you want to make the shopping experience better at grocery stores, ask a handful of customers what problems they face while shopping in a grocery store. Ask this question in several ways and record their answers. Continue the research online by searching for customers' problems at stores. You may do a Twitter #hashtag search on grocery chains such as #Kroger or #Publix to identify conversations people may be having and frustrations they may be experiencing. This can take some time, but the innovator can find several nuggets of insight to guide the future direction of the product and start-up business, while reducing the chance of business failure, which is the purpose of this exercise.

After gathering some potential list of customer problems or frustrations, test these formally with a larger audience. Get input from thirty or more people in your refined target audience by doing a Facebook survey of your friends (least expensive) or by using an Internet site such as "Ask Your Target Market" (costs several hundred dollars) to ask questions to an online panel of respondents you think could be your target audience.

During this learning and listening stage, wise advice to inventors and entrepreneurs is to *fall in love with the problem you are solving rather than a perceived solution that is dear to you*. Therefore, remain open to what you are learning from the target audience and be open to changing or discarding your pet solutions that do not find support.

Table 8.1. *Sample Survey to Identify the Significance of the Problem*

For a grocery store: A problem identification survey could look like this:

How big are these problems?	0	1	2	3	4	5	6	7	8	9	10
Finding my way around the store											
Getting a shopping cart WITHOUT wobbly wheels											
Finding product information											
Finding new products in the store											

Scale: Very small problem = 0; Very big problem = 10

Table 8.2. *Frequency of the problem*

How frequent are these problems	0	1	2	3	4	5	6	7	8	9	10
Finding my way around the store											
Getting a shopping cart WITHOUT wobbly wheels											
Finding product information											
Finding new products in the store											

Scale: Occurs very rarely = 0; Occurs very frequently = 10

Demographic or business-related information can be added to the survey, such as gender, age, location, income, and frequency of shopping, to define more precisely the target audience among your responders.

Once the survey responses are in, analyze the data to determine which problem statements are not significant and do not happen frequently; the product and start-up must not focus on these problems. Plot the significance of the problem on the x-axis and the frequency of the problem on the y-axis. The problem statements farthest from the origin are the ones deserving your focus and attention.

Developing and Testing Value Propositions

A value proposition says why an ideal customer would do business with you instead of your competitors. It explains why a customer would care and what your proposed solution does for customers. According to Osterwalder and Pigneur (2010), a value proposition addresses a pain a customer encounters, a gain or improvement sought, or solves a task for a customer. Osterwalder is best known for his one-page Business Model Canvas that helps potential entrepreneurs of future start-ups visually develop the nine components of a business model and test their assumptions using quick and inexpensive experiments. Before conducting a market survey regarding the benefits of a product or start-up, clearly define the value proposition that a product or start-up has to offer.

Grocery Example: Testing of Value Propositions

Going back to our grocery store example, a potential innovative idea could be locating products on the shelves of grocery stores using a smartphone. Say you developed a smartphone app that could help customers find items in unfamiliar stores. To test the value propositions for this app, you need to articulate one or more value propositions or benefit statements. Let's say one of many value proposition for the app is "Find any item in any grocery store in half the time using the Findit App." Initially, conduct a quick informal survey of 10 people. Read them the problem statement and your solution. If a significant majority does not think this idea solves a problem or provides value, then your idea needs reworking. The purpose of this quick, early stage test is to see if your idea has "legs." Assuming the quick test brings a positive response (7 or more out of 10 like it), use a slightly more formalized market survey. The methodology presented below was developed by the Eureka Ranch, a product and research development group that studies the introduction of innovative products and serves as consultant for major corporations to help them identify new and big ideas. Eureka Ranch found that there is a correlation between new product success and the sum of two weighted questions (see below) asked early on in the idea development process.

Eureka Ranch's approach takes the start-up idea's value proposition and asks thirty or more people two questions as part of a simple survey that can be conducted using paper and pencil, Survey Monkey, Ask Your Target Market, or other survey-based tools. The two questions to ask are: How "New and different" is this idea? And "How much do You Like it" at this price? The example of questions:

Scale: I really DISLIKE it = 0; I really LIKE it = 10

Q1. I like the solution to the problem at this price	0	1	2	3	4	5	6	7	8	9	10
App that helps you find items in a grocery store in half the time.											

Scale: NOT very different at all = 0; extremely new and different = 10

Q2. The solution is new or different	0	1	2	3	4	5	6	7	8	9	10
App that helps you find items in a grocery store in half the time.											

Eureka Ranch recommends, **for each survey respondent**, compute a "New and Different" rating – multiply response to Q2 by 0.4; and compute a "Like and Dislike" rating – multiply response to Q1 by 0.6. Add the two ratings together to find a **composite number** for the respondent. For example, if one respondent gives a score of 7 to the "New and Different" question, and an 8 for the "Like/Dislike" question; this respondent would get the following composite score: 7 x (.4) + 8 x (.6) = 7.6. Add the composite score for all respondents in your survey and divide by the total number of participants to get the start-up idea's **average composite score**. Eureka Ranch's rating table below tells the inventor what the next step should be; in the above example, the composite total of 7.6 merits a "wow" customer-response rating in the Eureka

Eureka Ranch's customer-response rating

Decision or Conclusion:	Average Composite Score
NO way!	0 to 3.8
RE-THINK	3.8 to 5.5
GO forward	5.5 to 7.2
WOW! Do it	7.2 to 10

Ranch's Customer-Response Rating table above, meaning the innovator may go ahead with the idea's development and additional testing.

Expanded Surveys

After obtaining responses from thirty or more people with either a rating in the "Go forward" or "Wow!" category, the inventor could go on to refine the target market and the value proposition using larger samples. Today, the Internet makes the execution of more traditional market research involving hundreds or more prospects easy to do. For example, an innovator could do online testing using a landing page linked to select Google AdWords Keywords. The use of AdWords is a great way to test a short value proposition on a limited budget. Alternatively, entrepreneurs can use a site such as QuickMVP to validate a value proposition on the Internet. MVP stands for Minimally Viable Product that can range from an idea's hand drawing or a computer-generated mockup of a product idea to an actual product with minimum features. The QUICKMVP.com site claims that within five minutes you can create a landing page that details your value proposition, select keywords to validate your idea.

Summary

The basic premise of this chapter is that you need early stage, inexpensive research to determine if your idea has "legs" before spending more time and money on formal studies and product development. Early on in the idea process, verify that you are addressing a real and hopefully frequent problem your audience has. Next, find out how new and different your product/start-up idea is to a selected target audience, as well as how much the audience likes or dislikes it.

Illustrative Case: Tennis Racquet Customer-Needs Survey

There are several textbooks on step-wise product development. Most of them, including the book by Ulrich and Eppinger (2007),[1] would suggest the following steps in the product development process:

1. Identifying Customer Needs
2. Product Specifications
3. Concept Generation
4. Concept Selection
5. Industrial Design for customer appeal, ergonomic design, etc.
6. Design for Manufacturing/scaling up production, efficiency in production
7. Prototyping.

The Tennis Racquet Case

In the tennis racquet case, a new entrant to the business of making and selling tennis rackets would need customer input through a survey of potential customers. Before a survey questionnaire could be developed, the following must be resolved:

1. Who to survey?
   a. If the racquets were to be made for professional tennis players, then survey professional players.
   b. If the target market is the local club–level recreational players, survey them.
2. What to gather from the survey?
   a. The racquet properties desired by the potential users;
   b. The properties that are missing in the racquets in the market;
   c. The price that potential customers are willing to pay;
   d. The preferred racquet weight;
   e. Other?
3. How many people to include in the survey?

---

[1] Karl Ulrich and Steven Eppinger, *Product Design and Development.* New York: McGraw-Hill/Irwin, 2007.

Number of Respondents to Your Survey

Survey Reliability: The number of people surveyed and the number who responded to your survey will determine the reliability of the survey findings (reliability tells you if you can depend on your survey findings – that is, if the results would hold good if it is repeated with a similar audience). A study based on a sample that is less than 30 could be less reliable than a survey based on 100 respondents – the more the better. It is better to avoid a sample that is less than 30 as a rule of thumb. For more on valid and reliable survey results, read books on survey methods.

    Survey validity: A survey is valid if its findings are applicable to the target market of interest. For example, in the case of tennis racquets, the findings of a survey of 100 men aged 85 or more living in a nursing home is not applicable if your target market is young men and women aged between 15 and 35 years—this is an exaggerated example employed to show how the result of a survey could be invalid.

    The following is a sample questionnaire for a tennis-racquet maker interested in finding general customer needs (this example is not very targeted; one should add or subtract questions, as needed):

Table 9.1. *Suggested sample questions for a survey of tennis players*

Question or item	Response or choice
TENNIS RELATED	
Do you play tennis?	If you do not play tennis, and do not intend to play in the near future, skip the rest of the survey
How often do you play tennis?	Circle one: More than once a week; four times a month; once a month; and so on
If you start playing tennis, where would you look for a racquet?	Circle all that apply: the Internet; *Tennis Magazine*; sports equipment stores; and so on
If you are a beginner, how much would you spend on a new tennis racquet?	Circle one: $16–30; $31–50; $51–100; and so on
If you already play tennis, and if you are shopping for a racquet, how much would you pay for a new racquet?	Circle one: $31–50; $51-$100; $101-$200, and so on
Are you member of a tennis club?	Yes / No
If you are able to provide or show the potential customer a description of your product, pictures, or prototype, its properties, etc., the following questions are relevant	
Would you purchase this racquet?	Yes / No / Maybe
If yes, how much would you pay for this racquet?	Circle one: $15–30; 31–60; etc.
Based on the information you have, how likely are you to buy this racquet?	Circle one: 0–20% chance; 21–40% chance; 41–60%; and so on
How much did you pay for the last racquet you purchased? Skip if you never purchased one	Circle one: $15–30; 31–60; etc.
What do you like or dislike about this racquet?	I like: I dislike:

Demographics

Question or item	Response or choice
Where do you live?	State:
	City:
Male or Female	M / F
Age range	Circle one: 18–24, 25–24; etc.
Health	Circle one: Excellent, above average, etc.
Education	Circle one: No high-school; high-school degree; two years of college; etc.
How often to do you play any sport?	Circle one: Twice a week; once a week; once a month; etc.
Annual household income	Circle one: <$25k; $26k – 50k; etc.
Do you own your home?	Yes / No
Market value of your home?	Circle one: <$100k; $101–250k; etc.

Based on the above survey's results, assuming it is valid and reliable, a customer needs statement and relative importance of customer needs can be prepared and given to the racket designer. See the following example of results from a hypothetical survey of 50 tennis players (you may rank the following results of the survey on relative importance for product design):

Hypothetical customer needs based on the survey:

1. Racquet to be sold through various outlets including Walmart for easy access to the new recreational player (rank …).
2. Price to be attractive to the noncompetitive new club-level player – under $40 (rank …).
3. The racquet to be capable of taking some abuse from distraught players who tend to throw the racquet (the survey in the table above could include a question to find out what percent of players are prone to do this) (rank …).
4. The racquet to have an appealing look for its owner to feel good about it (rank …).
5. The racquet to come with a shoulder case (rank …).

The above ranked list of items given to a racquet designer captures the needs of target customers gathered through the survey. In developing new product concepts for a tennis racquet for **advanced players**, the designer may get input from wider and more appropriate sources such as the following:

1. Consultations with club professionals, college players and coaches.
2. A study of competitors' advanced racquets and models.
3. A study of patents and patent applications on tennis racquets (via US Patents and Trademarks Office, or Google Patents, or other free databases for patents).
4. Tennis magazines, the Internet, and trade literature.

10 What Engineers Can Do for Product Development

What Engineers Can Offer

This chapter is a reminder to engineers about their core competence. This chapter also enables non-engineers working with engineers to get an idea of the core competence of engineers they work with.

Engineers learn to tame and master the laws and forces of nature in order to solve problems and meet our daily needs. For example, our need for travel in comfort and at great speeds may be defined as a problem or a need. Engineers have solved this for us by giving us the automobile, aircrafts, boats, and other fast-moving means of transportation. Engineers at the various auto-manufacturing companies have solved the problem uniquely to appeal to individual customers' needs; thus we have Mercedes-Benz, Ford, Fiat, Dodge, Toyota, and other companies solving our travel problems or needs with their cars and their features. Each car designed and built, employing sound engineering principles, overcomes human limits to travel.

A sample of what engineers from various disciplines could do is summarized in the list that follows. It is a partial overview of engineers' skillset.

1. **The ability to model** real-life complex problems or systems by engineers, either on paper or using mathematical models and the power of the computer, is a part of all engineers' skillset; such analyses enable their understanding of complex systems and make it possible to design new or improved problem solutions using new products/systems.
2. The ability to create **testing systems and procedures** for products for strength, performance, and **failure-proofing**. Examples:
   A. Aircrafts such as the Boeing 747, which can take off, fly, and land safely with a total weight of nearly one million pounds, without structural failure, are designed and tested by engineers.
   B. Bridges that do not collapse under the weight of vehicles crossing it, be it cars, trucks, or trains; engineers understand strength, vibration, failure, fracture, and other physical properties to produce strong structures.

3. The ability to harness **energy**, distribute it, store it, transfer it, and convert it in various ways, from electrical to mechanical, mechanical to electrical, electrical to chemical and so forth.
4. The ability to **generate large-scale power** from water, coal, nuclear, gas and other sources.
5. The ability to **control** large systems using microelectronic equipment/systems.
6. The ability to **substitute materials** for improved performance or for cost advantages.
7. The ability to improve **process efficiency and automation** to enable manufacturers and suppliers to become efficient and competitive.

Substitution

Engineers are constantly improving product performance and costs by the substitution of one engineering solution for another. Consider the following examples:

1. Aircraft controls were once entirely mechanical systems using levers, cables, pulleys, and hydraulics, whereas now they are almost entirely electrical; some call such planes "fly-by-wire planes," and most of today's planes fit this description. Purely mechanical engineering solutions in this application are giving way to partial electrical-engineering solutions.
2. Aircraft skin used to be entirely aluminum or aluminum alloy sheeting, which is now being replaced by carbon composites that not only reduce weight but also increase strength and reduce fuel consumption. Boeing 787, the newest in the Boeing fleet, uses carbon composites instead of aluminum; metallurgical engineering is giving way to chemical engineering in this product.
3. Tennis racquets have switched from wood (1970s) to metal, and from metal (1980s) to carbon composites, followed by many different compositions of the carbon composite including titanium, copper, basalt, nanoparticles, and so forth.
4. Automotive door windows used to have manual mechanical cranks to move the glass panes up or down. They used to be designed by mechanical engineers, but now they are almost always electrified; now electrical and mechanical engineers design car windows.

Product Improvement over Time

Engineers are always improving product designs to do more for less for the customer because of competition. If a product does not improve, the company becomes vulnerable to competitors, who can take away its markets/customers by introducing better products to attract customers from competitors.

Electronics and computers have many good examples of constant improvement by engineers. For example, a small laptop costing a few hundred dollars today has many, many times more computing power and storage capacity than large, mainframe computers forty years ago that cost hundreds of thousand dollars. Improvement to computer performance is the result of the contribution by electronic engineers, electrical engineers, mechanical engineers, and other engineering specialties.

In recent decades, explosive growth in microelectronics applications, polymer applications developed by chemical engineers, Internet applications, and biotechnology applications has created many new commercially successful products and businesses.

How to Design and Develop Better Products?

Products are designed by engineers for customer satisfaction through function, performance, and much more. Engineering design is a compromise among competing and conflicting design goals based on a list of customer needs.

Engineering Design Principles

1. Design for function incorporates in the product all the functions the customer expects from it.
2. Design for performance: design such that the product meets performance specifications and goals – for example, in the case of a car, it must be designed to deliver the horse power claimed, fuel efficiency claimed, the antilock brake system that works, airbags deploy on impact, etc.
3. Design for efficiency (minimize cost of input, fuel, energy, etc.).
4. Design for maintenance: low-cost maintenance, ease of access to the internals of the product to fix it; modular designs for quick and easy maintenance.
5. Design for strength or for failure-proofing: the product is not to fail when in use – a chair must not collapse under the weight of a seated person.
   A. Choose the right material and/or size.
   B. Use principles of engineering for load-bearing structures.
   C. Design with a factor of safety: design for more than the expected load – sometimes 50 percent to 100 percent more than the maximum expected load – but trade it off with the added cost.
6. Design for cost and affordability: use appropriate substitutes that are less expensive.
7. Design for manufacturability: a key aspect of design is to enable scalability to high-volume production, ease of manufacture, low-cost and high-quality manufacturing.
8. Design for safety: the finished product must not injure users or observers of the product.
9. Design for durability: long lasting, or at least as long as expected by the customer.
10. Design for reliability: product must perform as intended again and again – a car is reliable if one can depend on it day after day to take the owner from point A to point B without failure.
11. Design for customer appeal – the soft side of design.
12. Design to compete – to do what the competition offers or more.
13. Design to innovate – use new ideas to improve the product.

PATENTING AND PROTECTING YOUR INTELLECTUAL PROPERTY

11 Intellectual Property, Patents, and Trade Secrets

During Phase 2, while an innovator conducts product development, it is likely some new inventions related to the idea/product may emerge. Inventions need appropriate protection under the law to enable the innovator to reap rewards from the invention by licensing it to other businesses or through a new start-up business. Further, during Phase 4, potential investors would like to see one or more patent protections for key technologies at the heart of the business.

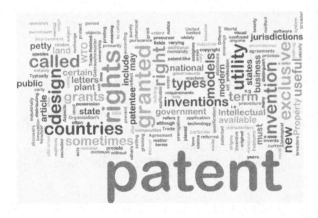

(© Kheng Guan Toh / Shutterstock, 75171478)

What Is a Patent?

In the United States, a patent is a legal, exclusive right to intellectual property (IP) for a period of twenty years. A patent is awarded to the inventor after careful examination by the US Patent and Trademark Office (USPTO), enabling the inventor to sue those who may infringe upon the issued patent claims during the time the patent is in force. Except when legally permitted (public utilities are examples of such exceptions), monopolies in commerce are illegal in the United States because they restrict competition; a patent is a temporary legally permitted monopolistic restriction on competition to allow the intellectual property holder (including licensees) to reap financial rewards from his or her invention for 20 years.

Patenting is the common form of IP protection for products or processes of technological innovation, but it requires complete disclosure of the invention through the patent application to the USPTO. Patent applications and all issued patents are published on the web by the USPTO.

Some selected patents are included in Chapter 13. You can also access and peruse various patents at several free websites including Google Patents (www.google.com/googlepatents/about.html), or the USPTO website (www.uspto.gov).

Types of Patents (According to the USPTO)

Among the patents granted by the USPTO, Utility Patents are the most common. In addition to Utility Patents, patent protection is available for (1) ornamental design of an article of manufacture (Design Patents) or (2) asexually reproduced plant varieties (Plant Patents). According to the USPTO, Utility Patents are granted for new, non-obvious and useful claims concerning:

- a process;
- a machine;
- an article of manufacture;
- a composition of matter, i.e., materials; and
- an improvement of any of the above.

What May Be Patented? (According to the USPTO)

To receive patent protection under the law, an invention must meet the following tests:

- novelty, newness, uniqueness, and non-obviousness;
- adequately described or enabled in the patent application (for someone with ordinary skills to make and use the invention); and
- claims by the inventor in clear and definite terms (NOTE: it is the claims that receive patent protection).

You can find claims of issued patents in Chapter 13 Supplements.

What Cannot Be Patented? (According to the USPTO)

According to the USPTO, the following cannot be patented:

a. laws of nature;
b. physical phenomena such as thunder, lightning, etc.;
c. abstract ideas;
d. literary, dramatic, musical, and artistic works (these are protected via copyright); and
e. inventions that are not useful, and are morally offensive to the public.

Provisional and Nonprovisional Utility Patents

A "Provisional" application for a Utility Patent could be filed first followed by a "Nonprovisional" Utility Patent application (for 20-year protection) within 365 days of the Provisional application; in doing so, the Provisional application date becomes the official filing date for the Nonprovisional patent application filed later. One can, however, file a Nonprovisional application without ever filing a Provisional one earlier.

In the "first to file" protocol adopted by the United States in 2012, the Provisional application gave up to 365 days advantage as well as IP protection for about $65 in filing fees for a micro-entity (low-income individuals, students) in 2015. A copy of the Provisional US patent application cover sheet can be found in Appendix C.

Patent Search

The term "patent search" describes the process used by legal and IP professionals to find similar issued patents or applications for patents before preparing a new patent application. The purpose is to ensure that no application is prepared for an invention that is already patented, or has a patent application pending for decision by the USPTO (as in a "patent pending" stage). Because of the cost and time associated with patenting an invention, it makes sense to search the publicly available patents and patent applications on the USPTO website[1] or on Google Patents[2] or other patent-databases[3] before applying for a patent. *A thorough patent search may convince the inventor to abandon the patent application, or teach the inventor to develop the invention to a patentable condition by making the patent unique, novel, and not infringing on existing patents or applications for patents.*

One can search the patents using verbal descriptions or using USPTO's patents classification numbers; you can find these numbers on the first page of all issued patents accessible on the web through one of the databases listed earlier. Resources for patent searches may be found in several US universities – example, http://www.lib.auburn.edu/find/bysubject.php#9, and the USPTO.

What Do You Get with an Issued Patent?

An issued nonprovisional utility patent has many sections; one of them is the Claims section. The legally enforceable patent rights granted to an inventor are contained in this section. They should be written carefully to avoid overlapping with other issued patents, and must claim all that the inventor deserves to claim as a result of the invention.

---

[1] http://www.uspto.gov/patent or http://www.uspto.gov/patents-application-process/search-patents
[2] https://patents.google.com/
[3] InnovationQ plus of IEEE or World IP Organization (WIPO) or PIUG Multinational patent databases or LexisNexis TotalPatent

An issued Nonprovisional Utility Patent is a temporary exclusive legal right given to the patent owner for twenty years under the new US patent laws since 2012 (used to be seventeen years). To enforce the patent, an issued patent gives the right to sue anyone alleged to have infringed on the patent. If the owner of the patent rights is not motivated to legally enforce the claims of the patent, the patent rights could become worthless to the patent owner.

When to File a Patent Application?

Until 2012, the US patent law used to be different from other nations' laws by using the "first to invent" rule for granting patents. As of 2012, USPTO is required by law to grant patents under the "first to file" rule. The implication is that if an inventor wants to protect his/her invention, he/she must apply for a patent earlier than others, or as soon as possible. Disclosing an unpatented idea to others before it is filed as a patent application may cause others to file a patent for a similar idea or an improved version of the idea. Organizations use a "non-disclosure agreement" (NDA) when discussing their unpatented or unprotected ideas with others; Auburn University's Secrecy/Non-Disclosure Agreement is provided as an example in Appendix B.

What Can You Learn from Issued Patents?

In short – a lot. Reading and comparing many issued patents can help an inventor understand the requirements for issued patents as well as the language used in a patent as a whole and in the Claims section in particular.

Trade Secret

Companies that own unique processes may not apply for patents for their processes in order to keep the trade secret away from their competitors. Superior processes provide cost, quality or flexibility advantages. Any practical idea that a company does not divulge through a patent is called a trade secret. A trade secret is only as good as the company's ability to keep it as such. Employees of a company using trade secrets are sworn to secrecy as a condition of their employment; they also may be prevented via an employment agreement from working for a competitor for a certain number of years after leaving the current employer, and may be prohibited from sharing the trade secret even after leaving the employer.

The best-known example of successful trade secrets is that of Coca-Cola ingredients (or formula), which has been kept a trade secret for more than 100 years. If it were patented, the Coca-Cola formula would have been disclosed to the whole world as required by the US patent law in return for legal monopoly for the first seventeen years after filing. Thus, it is "possible" to do well without patenting, but instances of such success are the exceptions rather than the rule. There are thousands of companies that have made IP protection through patents a successful strategy to get and keep customers. Nike, Microsoft, and Apple have patents for everyday products to give them competitive advantage.

What Can You Do with an Issued Patent?

The owner of an issued patent (or a patent application to the USPTO) may license the patent or potential patent to a licensee for a fixed amount, or for a periodic royalty payment for many years as a percent of sales tied to the invention. The other option is to start a new business with the new invention at the core of it. If a business uses the invention that is in the application stage without a granted patent, it is common to use the term "patent pending" with the product and associated documents to alert everyone that the IP is protected, and any violations of patent rights may be prosecuted.

International or Worldwide Patent Rights

Generally, US patent protection does not extend outside the United States, and vice versa. Therefore, an inventor in the United States must submit an application for patent rights in specific nations for IP protection outside the United States in addition to the application to the USPTO; the same is true for inventors in other nations wanting IP rights in the United States for their patents – they must submit an application to the USPTO.

The World Intellectual Property Organization (WIPO) and the Patent Cooperation Treaty (PCT) govern patent rights across international borders.[4] If an inventor in the USA wants patent protection outside the United States, the following are relevant:

1. Apply to the USPTO and then apply in the nation of interest outside the United States within the time limit imposed by the PCT. If interested in international patent rights, make an additional application under PCT to the USPTO, when you apply for US patent rights. USPTO can advise you over the phone on the process of applying under PCT for international patent rights.
2. The cost of protecting IP rights in multiple nations could be prohibitive to an individual inventor or to a small company; in such circumstances, one could apply for a US patent in a timely manner and find interested large firms that may license the technology in the United States and also apply in foreign nations for international patent rights.
3. If a US-patented invention is not patented in other countries as well, it may be infringed upon outside the United States without legal recourse. However, a US-patent-infringing product made outside the United States by a company that in not based in the United States can be legally prevented from entry into the US market using the rights granted under the US patent; therefore,

---

[4] http://www.wipo.int/about-wipo/en/; and http://www.wipo.int/pct/en/texts/articles/atoc.htm

international firms wanting to enter the large US markets legally must honor US patent rights. Thus, the enforcement of patent rights in the US extends to foreign firms wanting to operate in the US market.

International patent rights are sometimes difficult to enforce. Even Microsoft, a powerful international player, finds it very difficult to enforce the rights for its software in certain countries where a culture of patent infringement prevails.

12 If You Cannot Afford a Patent Attorney

The Cost of Getting a Patent: Should You Apply for a Patent?

A note of caution to inventors: around 90 percent of issued patents do not produce any income to the inventor because they are either not licensed to a royalty-paying commercial business or are not part of a start-up business created by the inventor. Therefore, while inventions are encouraged, the cost of patenting must be weighed against potential returns from the invention to cover the cost of the invention.

If an inventor has several patentable ideas and does not have the financial capability to engage an attorney, such an inventor is encouraged – as well as given guidance, in the later chapters – to file for patents without using an attorney for the most promising of their inventions with prospects of licensing or a start-up.

If one cannot afford the services of a patent attorney, it is better to file a patent application as a pro se (that is, not represented by an attorney) applicant at a much lower expense. The result may not be a perfect patent sometimes, but it is likely to be better than no patent at all if the inventor can license the product or use it to bring about a start-up business.

Some self-education about patenting can prepare a pro se applicant. The inventor is encouraged to search the millions of published patents and patent applications on the USPTO website and/or Google Patents, and become educated and savvy before writing the application and the claims; patent applications must follow the same format and pattern as recently published patents.

The simplest of inventions filed by an individual inventor (called small entity) in the United States, without any attorney services, may cost about $600–700 (2015 prices) in application fees, and another fee of about $1,300 at issue. If one has income low enough to be classified as a micro-entity by the USPTO – almost all college students, for example, would qualify as micro-entities – these costs will be reduced in half.

Table 12.1. *Independent inventors granted utility patents by USPTO*

Inventions from:	US total	1975–1997	1998–2014	2014	Average/ year since 1998	State pop. (million)	2014 inventions per mil pop.
Independents	605,355	307,638	297,717	19,377	17,512	318.8	54.9
California		57,358	60,857	4,237	3,579	38.8	109.2
Alabama		2,099	2,218	118	130	4.8	24.6
Florida				1,434		19.9	72.1
Texas				1,277		26.9	47.5
New York				1,257		19.7	63.8

Source: Population July 1, 2014 US Census Bureau; http://www.uspto.gov/web/offices/ac/ido/oeip/taf/inv_utl.htm

Individual Inventor, the Pro Se Applicant

The US government has made it possible and easier for the individual inventor to apply as a pro se applicant (without a patent attorney) and get patent protection.

US Patents and Trademarks Office (USPTO) has thousands of examiners who study each patent application before either rejecting or granting it. They are qualified engineers/scientists. The USPTO website requires this qualification of examiners of engineering content in patent applications: "Successful completion of a full 4-year professional engineering curriculum leading to a bachelor's degree or higher degree in engineering in an accredited college or university." Therefore, an engineer-inventor is qualified to serve as an USPTO examiner after the training provided by USPTO.

Table 12.1 indicates that 605,355 utility patents were granted to individuals in the United States since 1975; 19,377 of them were issued in 2014. In the last column we see there were 109 patents issued to independents for every million residents in the state of California; in Alabama, 25 patents were issued to independent inventors for every million residents in the state; i.e. more individual inventors are active and patenting in California than Alabama.

Table 12.2 shows that in 2014 there were 578,802 total applications for utility patents in the United States, and 300,678 – about 52 percent – were granted. It should be noted that patent applications may take one to four years to be granted.

Since 19,377 patents were granted to independents in 2014, this means that 6.4 percent of all granted utility patents in 2014 belonged to independent inventors; the other 93.6 percent were granted to corporations, business entities, or nonprofit entities, which can afford a patent attorney. Only independent inventors are likely to apply as pro se, without the services of a patent attorney because of affordability issues.

USPTO's Assistance to Pro Se Applicants

Over the years, USPTO has made many improvements to the patenting system to help individual inventors, many of whom are pro se applicants, applying without the services of a patent attorney. Several helpful USPTO sites are listed in the section

Table 12.2. *All utility patent applications and granted patents, United States*

All Inventions	Applications US	Applications foreign	Total appl.	Granted	% granted
All US 2014	285,096	293,706	578,802	300,678	52.0
2013	287,831	283,781	571,612	277,835	48.6
2012	268,782	274,033	542,815	253,155	46.6

Source: http://www.uspto.gov/web/offices/ac/ido/oeip/taf/us_stat.htm

on USPTO Resources that follows. *These resources are mostly aimed at individual inventors and pro se applicants.* Notable resources are:

1. The Pro Bono Program, which provides free patent attorney services to qualified inventors who cannot afford a patent attorney – contact the USPTO for information: Item 5 below.
2. The *Inventors Eye* Newsletter, which informs the independent inventor of all the new services and developments relevant to the independent inventor.
3. The video on Provisional Patent application: Item 11 below.

United Inventors Association (www.uiausa.org) is a US nonprofit organization helping independent inventors and pro se patent applicants. Their website has numerous contacts and avenues for patenting and for commercializing inventions by independents. They help protect and promote the rights of the independent inventor.

USPTO Resources

1. USPTO, 16th Annual Independent Inventors Conference, August 15–16, 2014, http://www.uspto.gov/about-us/events/16th-annual-independent-inventors-conference
2. USPTO, Checklist for Filing a Nonprovisional Utility Patent Application, http://www.uspto.gov/sites/default/files/inventors/Checklist_for_Filing_a_Nonprovisional_Utility.pdf
3. USPTO, US Patent Classification System, http://www.uspto.gov/page/examiner-handbook-us-patent-classification-system
4. USPTO, Patent "Lunch and Learn," November 19, 2015, Silicon Valley USPTO, San Jose, CA, http://www.uspto.gov/about-us/events/patent-lunch-and-learn
5. USPTO, Patent Pro Bono Program, http://www.uspto.gov/patents-getting-started/using-legal-services/pro-bono/inventors
6. USPTO, Published Complaints, http://www.uspto.gov/patents-getting-started/using-legal-services/scam-prevention/published-complaints
7. USPTO, Independent inventors by state by year: Utility Patents Report, January 1975–December 2014, http://www.uspto.gov/web/offices/ac/ido/oeip/taf/inv_utl.htm
8. USPTO, Inventors & Entrepreneurs Resources, http://www.uspto.gov/learning-and-resources/inventors-entrepreneurs-resources
9. USPTO, *Inventors Eye* Newsletter, http://www.uspto.gov/learning-and-resources/inventors-eye-newsletter

10. USPTO Provisional Application for Patent, http://www.uspto.gov/patents-getting-started/patent-basics/types-patent-applications/provisional-application-patent
11. USPTO video on provisional patent application, https://www.youtube.com/watch?v=qIGrAg5AVL8
12. USPTO video for submitting a provisional application electronically, https://www.youtube.com/watch?v=B--8oFg88Uw
13. USPTO, Webinar for Patent Pro Se, Pro Bono, and Law School Clinic Certification Programs, http://www.uspto.gov/sites/default/files/documents/ProBono_ProSe_Law_School_Clinic_Certification_Program.pdf

CHAPTER 12 SUPPLEMENT 1

USPTO Establishes Special Examination Unit for Pro Se Applicants

http://www.uspto.gov/blog/director/entry/uspto_establishes_special_examination_unit
Monday, Nov. 24, 2014.

Blog by Deputy under Secretary of Commerce for Intellectual Property and Deputy Director of the USPTO Michelle K. Lee

(Partly reproduced below)

"The United States Patent and Trademark Office (USPTO) is committed to assisting inventors by offering education and tools to those who file US patent applications without the help of a patent attorney or agent. This is known as pro se filing. Our Office of Innovation Development (OID) has a long history of helping pro se filers and independent inventors understand and navigate the patenting process as well as offering a **variety of resources** and outreach programs to the public. The agency is now expanding these services by piloting a special unit focused on examining applications filed by pro se applicants. The new Pro Se Pilot Examination Unit is the product of an **executive action** issued by the White House earlier this year.

The Pro Se Pilot Examination Unit began in October 2014 and will operate for at least one year. Comprised of experienced examiners from all scientific disciplines, these examiners receive training surrounding issues often encountered by pro se applicants, such as how to respond to a Notice of Missing Parts or how to revive an unintentionally abandoned application. In addition, the examiners provide **customer support** and answer general patent-related questions via a toll-free number, email, or a walk-in service. Lastly, they spearhead development of specialized training materials on the intricacies of filing a patent application."

CHAPTER 12 SUPPLEMENT 2

Low-Cost Provisional Patent Applications

Why Do You Need a Provisional Patent?

USPTO offers a low-cost patent protection for everyone. For a college student, the cost of filing was around $65 in 2015 as a micro-entity; you need to apply to the

USPTO using the form included in Appendix C3 to be classified as a micro-entity based on your income. Despite the low cost, it gives the inventor a valuable priority date of filing and prevents others from stealing the idea/product being patented. This priority date is good for 365 days. During this time, the inventor can file a Nonprovisional Utility patent (for 20-year protection) as a pro se or with the services of a patent attorney. Provisional patent applications are very simple to prepare and submit for a number of reasons:

1. No rigorous Claims statements are required, unlike for the Nonprovisional application.
2. No oath or declaration is required, again unlike for the Nonprovisional application.
3. A filing date will be granted to a Provisional application when it contains a clear written description of the invention (details in **35 U.S.C. §112**(a)).
4. Provisional application is not examined by the USPTO and does not have to conform to strict guidelines; USPTO merely grants the date of application and files the application without examination.

What Do You Get for a Small Fee?

The small fee for a Provisional patent application is deceptive because the benefits are substantial for a year. It does, however, allow the inventor to move forward and start making and selling a product under the protection offered by the "patent pending" status.

1. It establishes an early effective filing date for a Nonprovisional Utility application to follow; Nonprovisional application must reference to the earlier-submitted Provisional application.
2. It allows the inventor the authorized use of the "patent pending" status to move ahead with commercialization activities such as manufacturing, marketing, and so forth.
3. It gives 12 months of IP protection, which can be extended under certain conditions.
4. It gives protection against stealing of the invention.
5. It ties the name of all inventors to the invention.

Amendments are not permitted in Provisional applications after filing, other than those to make the provisional application comply with applicable regulations.

Helpful USPTO Contacts for Applicants

The Inventors Assistance Center (IAC) of the USPTO provides patent information and services to the public. The IAC is staffed by former Supervisory Patent Examiners and experienced Primary Examiners who answer general questions concerning patent examining policy and procedure. See www.uspto.gov/inventors/iac/index.jsp.

CHAPTER 12 SUPPLEMENT 3

Low-Cost Nonprovisional Patents

Protecting Proprietary Ideas in Phase 1

If your home is worth $100,000 to $500,000, you protect it by insuring it against fire and other hazards, and by obtaining a title to the property. Similarly, when an engineer/scientist creates value during Invention Phases 1 and 2, the idea or product needs the protection offered by a patent. An innovator's proprietary idea is an intellectual property from a legal and commercial point of view. If the idea were to become the source of an income stream, and if it is a worthy and appealing idea, it is likely to be imitated by competitors, who can take your customers away from your business and cause it to fail.

The US law allows an inventor to seek legal rights and protection to intellectual property so that the idea can be commercialized under a limited and temporary monopoly rights granted under the US patenting system – 20 years for a Nonprovisional utility patent.

However, the cost of patenting through attorneys may be a barrier to many young inventors. The US Patent and Trademark Office (USPTO) allows individuals to apply for patents without using a patent attorney. The USPTO offers assistance over the phone to independent inventors and pro se applicants. Additionally, the USPTO provides free copies of its newsletter *Inventors Eye* for the benefit of the independent inventor.[1]

Nonprovisional Patenting Fees Are Affordable

The USPTO patenting fees for a Nonprovisional Utility patent are determined on the basis of income and type of applicant. They fall into three categories: the standard, small entity, and micro-entity. Small entities pay one-half of the fees paid by standard entities (businesses), while micro-entities pay one-half of what small entities pay. Most college students would qualify as a micro-entity; they would have to submit a completed USPTO form for micro-entity classification along with their Provisional or Nonprovisional patent application; copy of the form can be found in Appendix C3.

For micro-entities not using the services of a patent attorney the total cost for a granted Nonprovisional Utility patent good for 20 years for a relatively simple invention is about $1,000 to cover USPTO fees (2015 pricing). USPTO fees double for a small entity.

---

[1] USPTO, Inventor's Eye Newsletter, http://www.uspto.gov/inventors/independent/eye/201404/inventorseye_new_orgs.jsp

In contrast, an average cost for using the services of an attorney for the application process is about $10,000 per patent – beyond the reach of most college students and young inventors. On account of the cost, for young engineer/scientist-entrepreneurs and college students, pro se patent application is the most viable – and thus most recommended –option. The published articles of the author – included in this book – based on his experience as a pro se applicant may guide the reader interested in filing as a pro se applicant.[2]

Conditions for a Nonprovisional Application (Twenty-Year Protection)

You must include the appropriate filing fee disclosed on the USPTO website on the date of filing. Include as many drawings as possible with the application for the thorough understanding of the invention by the examiner (details in 35 U.S.C. 113). A *new drawing* necessary to understand the invention cannot be introduced into an application after the filing date because of the prohibition against new matter. Guidelines for the successful filing of a Nonprovisional application are:

1. Disclosure of the invention must be as complete as possible.
2. Each inventor must be named in the application.
3. All inventor(s) named in the application must have made a contribution to the invention.
4. The Nonprovisional application must have at least one inventor in common with the inventor(s) named in the Provisional application to claim benefit of the Provisional application filing date.

Warning from the USPTO site: "Beware that an applicant who publicly discloses his or her invention (e.g., publishes, uses, sells, or otherwise makes available to the public) during the 12 month Provisional application pendency period may lose more than the benefit of the Provisional application filing date if the 12 month Provisional application pendency period expires before a corresponding Nonprovisional application is filed. Such an applicant may also lose the right to ever patent the invention" that is described in the Provisional patent application.

Therefore, file the Nonprovisional application before the 12 month (365 days) anniversary of the Provisional application. Remember: If you do not file the Nonprovisional application in 12 months after the Provisional application, *and* if you have publicly disclosed the invention, you may lose the right to patent the invention described in the Provisional application.

Once you have a patent application on file with the US patent office, you have the option of looking for businesses that would license it in return for royalty income to you, or you may consider starting a business.

---

[2] Paul Swamidass, "Does the patent office snub inventors without an attorney," *Inventors Digest*, April 2010; http://idmagazine.wpengine.com/articles/does-the-patent-office-snub-inventors-who-file-without-an-attorney/

No Need for Prototypes Prior to Patenting

Frequently, new inventors and students interested in applying for a patent assume that a working prototype is necessary before they submit an application for a patent in the United States. *If the invention can be presented clearly in words and drawings for others to understand and use it, it can be patented without a working prototype.*

CHAPTER 12 SUPPLEMENT 4

USPTO Examiners Help Nonprovisional Pro Se Applicants

Inventor-Filed Applications

The Manual of Patent Examination Practices (MPEP) used by patent examiners of the USPTO to examine Nonprovisional utility patent applications explicitly instructs USPTO examiners to assist pro se applicants to obtain a granted patent if their application does not meet the requirements of USPTO but has meritorious content worthy of a patent.

MPEP Directive to Examiners to Help Inventors

Section 707.07(j) of MPEP is individual inventor–friendly. Every pro se patent applicant must read the paragraphs below carefully and thoroughly to understand the extent of help available to applicants without a patent attorney. The following are quoted from this inventor-friendly section, where three items are extremely encouraging to inventors who file their own patent applications.

1. "When, during the examination of a *pro se* application it becomes apparent to the examiner that there is patentable subject matter disclosed in the application, *the examiner should draft one or more claims for the applicant and indicate in his or her action that such claims would be allowed if incorporated in the application by amendment.* This practice will expedite prosecution and offer a service to individual inventors not represented by a registered patent attorney or agent. Although this practice may be desirable and is permissible in any case deemed appropriate by the examiner, *it is especially useful in all cases where it is apparent that the applicant is unfamiliar with the proper preparation and prosecution of patent applications.*"
2. "When an application discloses patentable subject matter and it is apparent from the claims and applicant's arguments that the claims are intended to be directed to such patentable subject matter, but the claims in their present form cannot be allowed because of defects in form or omission of a limitation, the examiner should not stop with a bare objection or rejection of the claims. The examiner's action should be constructive in nature and, when possible, should offer a definite suggestion for correction. Furthermore, an examiner's suggestion

of allowable subject matter may justify indicating the possible desirability of an interview to accelerate early agreement on allowable claims. If the examiner is satisfied after the search has been completed that patentable subject matter has been disclosed and the record indicates that the applicant intends to claim such subject matter, the *examiner may note in the Office action that certain aspects or features of the patentable invention have not been claimed and that if properly claimed such claims may be given favorable consideration.* If a claim is otherwise allowable but is dependent on a canceled claim or on a rejected claim, *the Office action should state that the claim would be allowable if rewritten in independent form.*"

3. "*Where the examiner is satisfied that the prior art has been fully developed and some of the claims are clearly allowable, the allowance of such claims should not be delayed.*" (All three items quoted from MPEP 707.07(j); emphasis added.)

The author has applied as a pro se and was granted four patents by the USPTO between 2009 and 2014. In each application the author requested in writing that the patent examination be subject to the provisions of MPEP 707.07(j) listed above.

Based on the author's multiple successful Nonprovisional patent applications, the most notable recommendations to patent applicants without attorneys are:

1. Prepare the Abstract, Specifications, and Drawings sections of the application as well as possible using published patents on a similar invention as a guide. Be thorough and convincing; these sections are likely to be accepted without any change, as was the case with all of the author's applications; well-written sections show to the examiner that you are familiar with and understand patent application format and contents.

2. Write the claims after reading many granted patents on similar inventions; your claims may be rejected, but examiners will rewrite them for you if there is patentable matter. This worked for me every time.

Reading and Learning from a Granted
US Patent

How to Read a Granted US Patent?

A "granted" US patent is one that has gone through a valid application phase, successful examination by the examiners of the patent office, approved and granted by the USPTO, with all the fees associated with the examination and publication of the patent promptly paid. Such granted utility patents give exclusive rights to the inventor(s) for twenty years.

In the following pages, in Supplement 3 of this chapter, US patent 6,376,126 is reproduced in its original form. It can be used for careful study during a patent search that could lead to licensing or outright purchase from the owner of the patent. Moreover, pro se inventor-applicant must read and learn from published patents in the area of his/her invention before drafting a patent application. The effort devoted to studying granted patents before preparing a pro se patent application for a Provisional or Nonprovisional Utility US patent will pay off in the form of a granted patent, and/or reduce the time spent in prosecuting the application. "Prosecution" refers to the effort of the inventor or his/her agents to respond to the USPTO questions and actions, and to convince the USPTO that a patent is deserved. The important information gathered from the first page of the granted patent in Figure 13.1 are:

1. Patent number   US 6,376,126
2. Date patent was granted   April 2, 2002
3. Inventor   Bernard Rivkin
4. Date of application   March 23, 2000
5. Application number   09/535,082
6. Address of the inventor – useful for approaching the inventor for licensing or purchase of the invention
7. International and US classification of the product – useful for patent search on similar products
8. Classifications under which patent searches were conducted by the USPTO prior to awarding the patent to the inventor – a new inventor with a similar

product would be wise to search these US classifications to see the prior art and to ensure that his/her idea is not patented:

a. US Classifications listed are: 24/10 R, 3.3, 303; 24/66.1, 3.8, 114.6;

9. References to other issued patents are cited. These patents may have been cited by the inventor or the patent examiner during the examination of this application. The patents on this list do not infringe on the new patent being issued; this is good to know for a new inventor while starting a patent search for a similar product, and he/she would be wise to go back and look up the patents on this list during his/ her patent search, and before writing a patent application for a similar invention.

10. If a foreign patent application was made, it will be recorded in this page.

11. If a patent attorney or law firm drafted and prosecuted the application, they will be identified in this page (Johnson and Stainbrook, LLP).

12. Finally, an Abstract of the patented invention appears on this page.

Figure 13.1. Page 1 of a granted Patent

US006367126B1

(12) **United States Patent**
Rivkin

(10) **Patent No.:** US 6,367,126 B1
(45) **Date of Patent:** Apr. 9, 2002

(54) **MAGNETIC FORCE EYEGLASS HOLDER**

(76) Inventor: **Bernard Rivkin,** 29 Oak Forest Pl., Santa Rosa, CA (US) 95409

(*) Notice: Subject to any disclaimer, the term of this patent is extended or adjusted under 35 U.S.C. 154(b) by 0 days.

(21) Appl. No.: **09/535,082**

(22) Filed: **Mar. 23, 2000**

(51) Int. Cl.⁷ **A44B 21/00**; A45F 5/02

(52) U.S. Cl. .... **24/3.3**; 24/10 R; 24/66.1; 24/114.6; 24/303

(58) Field of Search .... 24/10 R, 3.3, 303, 24/66.1, 3.8, 114.6

(56) **References Cited**

U.S. PATENT DOCUMENTS

2,319,292	A	*	5/1943	Boggs 24/303
2,363,914	A	*	11/1944	Wakefield 24/303
2,644,212	A	*	7/1953	Markowitz 24/303
3,129,477	A	*	4/1964	Mizuno 24/303
3,159,372	A	*	12/1964	McIntosh 24/303
3,161,932	A	*	12/1964	Russell 24/303
3,178,784	A	*	4/1965	Krauthamer .... 24/10 R
4,136,934	A		1/1979	Scron 351/157
4,264,821	A	*	4/1981	Bauer 250/480

4,458,384	A		7/1984	Arnold 24/3.3
4,894,887	A	*	1/1990	Ward, II 24/3.3
4,969,239	A	*	11/1990	Bruno 24/3.3
5,682,648	A	*	11/1997	Miller 24/303
5,682,653	A	*	11/1997	Berglof et al. 24/303
5,699,990	A		12/1997	Search 248/309
5,794,312	A		8/1998	O'Mahony 24/3.3
5,839,708	A		11/1998	Search 248/309
5,842,613	A		12/1998	White 24/3.1
5,845,369	A	*	12/1998	Dunchock 24/3.3
5,860,191	A		1/1999	Sieger 24/3.3
5,864,924	A		2/1999	Rodriguez 24/3.3
5,956,812	A		9/1999	Moennig 24/3

FOREIGN PATENT DOCUMENTS

GB		1293858	* 10/1972	.... 24/303

* cited by examiner

Primary Examiner—Victor N. Sakran
(74) *Attorney, Agent, or Firm*—Johnson & Stainbrook, LLP; Larry D. Johnson; Craig M. Stainbrook

(57) **ABSTRACT**

A method for holding eyeglasses using a magnetic force means in cooperation with a magnetically saturable keeper element whereby a convenient removable and reusable non-invasive securing system is created which may be used on apparel and other surfaces.

7 Claims, 4 Drawing Sheets

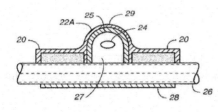

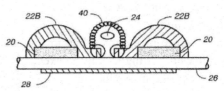

CHAPTER 13 SUPPLEMENT 1

US Patent: 7479949 "Touch Screen Device..." Steve Jobs and Others (Excerpts Only)

Touch screen device, method, and graphical user interface for determining[1] ...
Steven P. Jobs et al (3 pages of 362 pages shown here)

Patent number: 7479949
Filing date: Apr 11, 2008 **Issue date**: Jan 20, 2009
Application number: 12/101,832

Inventors:

Steven P. Jobs, Scott Forstall, Greg Christie, Stephen O. Lemay, Scott Herz, Marcel van Os, Bas Ording, Gregory Novick, Wayne C. Westerman, Imran Chaudhri, Patrick Lee Coffman, Kenneth Kocienda, Nitin K. Ganatra, Freddy Allen Anzures, Jeremy A. Wyld, Jeffrey Bush, Michael Matas, Paul D. Marcos, Charles J. Pisula, Virgil Scott King, Chris Blumenberg, Francisco Ryan Tolmasky, Richard Williamson, Andre M. J. Boule, Henri C. Lamiraux

Original Assignee: Apple Inc.
Primary Examiner: Duc Q Dinh
Attorney: Morgan, Lewis & Bockius LLP
Current US Classification: 345/173; 345/169; 715/784; 715/786

Abstract

A computer-implemented method for use in conjunction with a computing device with a touch screen display comprises: detecting one or more finger contacts with the touch screen display, applying one or more heuristics to the one or more finger contacts to determine a command for the device, and processing the command. The one or more heuristics comprise: a heuristic for determining that the one or more finger contacts correspond to a one-dimensional vertical screen scrolling command, a heuristic for determining that the one or more finger contacts correspond to a two-dimensional screen translation command, and a heuristic for determining that the one or more finger contacts correspond to a command to transition from displaying a respective item in a set of items to displaying a next item in the set of items.

Claims [Readers to note: Only one out of many claims shown; formatting reproduced exactly as shown in the patent]
Claim 1. A computing device, comprising:

> a touch screen display;
> one or more processors;
> memory; and
> one or more programs, wherein the one or more programs are stored in the memory and configured to be executed by the one or more processors, the one or more programs including:

---

[1]	http://www.google.com/patents?id=dCKzAAAAEBAJ&printsec=abstract&zoom=4#v=onepage&q&f=false

instructions for detecting one or more finger contacts with the touch screen display;

instructions for applying one or more heuristics to the one or more finger contacts to determine a command for the device; and

instructions for processing the command;

wherein the one or more heuristics comprise:

a vertical screen scrolling heuristic for determining that the one or more finger contacts correspond to a one-dimensional vertical screen scrolling command rather than a two-dimensional screen translation command based on an angle of initial movement of a finger contact with respect to the touch screen display;

a two-dimensional screen translation heuristic for determining that the one or more finger contacts correspond to the two-dimensional screen translation command rather than the one-dimensional vertical screen scrolling command based on the angle of initial movement of the finger contact with respect to the touch screen display; and

a next item heuristic for determining that the one or more finger contacts correspond to a command to transition from displaying a respective item in a set of items to displaying a next item in the set of items.

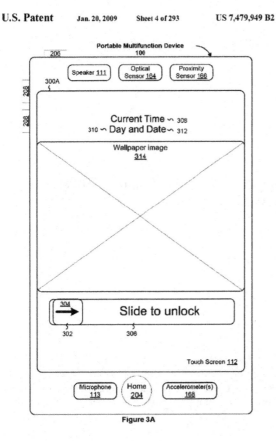

U.S. Patent   Jan. 20, 2009   Sheet 4 of 293   US 7,479,949 B2

Figure 3A

Figure 13.S1. One of the figures in the granted patent

NOTE: using the patent number above, the reader is encouraged to use Google Patents Search to find the entire patent published by the USPTO.

CHAPTER 13 SUPPLEMENT 2

ReadeREST: An Example of a Successful Patent

Earlier in this book, ReadeREST was introduced without showing the patent document behind this invention. In this chapter the full patent is included.

The *Shark Tank* TV show[2] has launched many successful products, but one of the most successful and yet simple inventions is ReadeREST. The patent document for this invention by Bernard Rivkin follows. This patent was commercialized when, years after the patent was granted, it was acquired by Rick Hopper for $5,000. Rick Hopper built the product in his garage and sold thousands before appearing on *Shark Tank*.

He asked for a $150,000 investment from the "sharks" in exchange for a 15% share of his company. He accepted an offer of $150,000 for 65% of the company from one of the sharks, Lori Greiner, called the "Queen of QVC TV show" for her success in launching hundreds of new as well as patented products on QVC. With her help ReadeREST reached about $10 million in sales in about two years, and sales continue to grow.

One of the examples of the many variations of the product now (2016) available on Amazon.Com for sale is in Figure 13.S2.

Figure 13.S2. One version of ReadeREST (Amazon.com)

---

[2] For more information, see *Shark Tank*, Season 3, Episode 302, November 30, 2012: http://sharktank-success.blogspot.com/2012/11/readerest-specsecure.html

A Complete Patent Document

The reader is encouraged to study the complete granted patent for ReadeREST on the next several pages. Read it carefully to understand what the major sections in the patent are, how each is written, and learn how to write similar sections for your patent application. (Note: even though the original patent application consisted of eighteen drawings, only the first eight are reproduced below.)

US006367126B1

(12) **United States Patent**
Rivkin

(10) **Patent No.:** **US 6,367,126 B1**
(45) **Date of Patent:** **Apr. 9, 2002**

(54) **MAGNETIC FORCE EYEGLASS HOLDER**

(76) Inventor: **Bernard Rivkin**, 29 Oak Forest Pl., Santa Rosa, CA (US) 95409

(*) Notice: Subject to any disclaimer, the term of this patent is extended or adjusted under 35 U.S.C. 154(b) by 0 days.

(21) Appl. No.: **09/535,082**

(22) Filed: **Mar. 23, 2000**

(51) Int. Cl.[7] A44B 21/00; A45F 5/02
(52) U.S. Cl. .... 24/3.3; 24/10 R; 24/66.1; 24/114.6; 24/303
(58) Field of Search .... 24/10 R, 3.3, 303, 24/66.1, 3.8, 114.6

(56) **References Cited**

U.S. PATENT DOCUMENTS

2,319,292 A	* 5/1943	Boggs	24/303
2,363,914 A	* 11/1944	Wakefield	24/303
2,644,212 A	* 7/1953	Markowitz	24/303
3,129,477 A	* 4/1964	Mizuno	24/303
3,159,372 A	* 12/1964	McIntosh	24/303
3,161,932 A	* 12/1964	Russell	24/303
3,178,784 A	* 4/1965	Krauthamer	24/10 R
4,136,934 A	1/1979	Seron	351/157
4,264,821 A	* 4/1981	Bauer	250/480

4,458,384 A	7/1984	Arnold	24/3.3
4,894,887 A	* 1/1990	Ward, II	24/3.3
4,969,239 A	* 11/1990	Bruno	24/3.3
5,682,648 A	* 11/1997	Miller	24/303
5,682,653 A	* 11/1997	Berglof et al.	24/303
5,699,990 A	12/1997	Search	248/309
5,794,312 A	8/1998	O'Mahony	24/3.3
5,839,708 A	11/1998	Search	248/309
5,842,613 A	12/1998	White	24/3.1
5,845,369 A	* 12/1998	Dunchock	24/3.3
5,860,191 A	1/1999	Sieger	24/3.3
5,864,924 A	2/1999	Rodriguez	24/3.3
5,956,812 A	9/1999	Moennig	24/3

FOREIGN PATENT DOCUMENTS

GB	1293858	* 10/1972		24/303

* cited by examiner

Primary Examiner—Victor N. Sakran
(74) *Attorney, Agent, or Firm*—Johnson & Stainbrook, LLP; Larry D. Johnson; Craig M. Stainbrook

(57) **ABSTRACT**

A method for holding eyeglasses using a magnetic force means in cooperation with a magnetically saturable keeper element whereby a convenient removable and reusable non-invasive securing system is created which may be used on apparel and other surfaces.

7 Claims, 4 Drawing Sheets

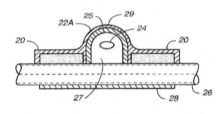

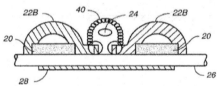

U.S. Patent   Apr. 9, 2002   Sheet 1 of 4   US 6,367,126 B1

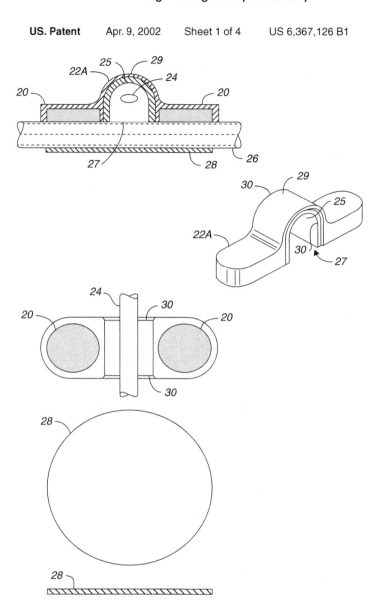

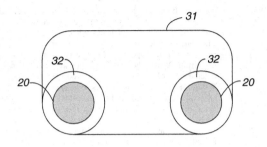

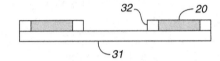

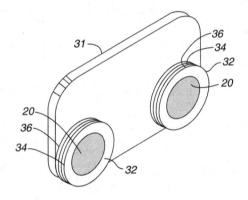

US 6,367,126 B1

1

MAGNETIC FORCE EYEGLASS HOLDER

BACKGROUND AND FIELD OF INVENTION

The invention is not a new material, but rather, the discovery of a new and novel solution to a common problem with the use of special materials.

Users of eyewear frequently wish to remove them temporarily. Now comes the problem of where to put them. If you have a pocket to put them in, what else is in the pocket that will scratch the lenses? When you want to retrieve the glasses, which pocket are they in? If you put them down on a table, will you put them down on the frames or on the lenses that may be damaged? Then will you remember to pick them up and not leave them behind? If you have sunglasses or other special glasses as well, that's a whole new set of problems.

The problems outlined above are currently being served in several ways. Cords or chains that attach to the eyeglass temples and are worn around the neck are the most typical present devices. Such devices are disclosed in U.S. Pat. No. 5,956,812 also U.S. Pat. No. 4,136,934 and many others. These are unsatisfactory for many reasons. Namely, when the glasses are being held they are awkward and uncomfortable, especially when the wearer is in motion. If the chain or cord is kept attached to the glasses, it is very uncomfortable while wearing. If you take the chain off and on it is a nuisance and a storage problem, especially if you need the glasses to find and install the temple attachments. Metal and mechanical holders that are fitted in the pocket are awkward and unattractive and tend to pull the pocket down because of the weight of the glasses and the attachment. Some examples are shown in U.S. Pat. Nos. 5,842,613, 56,999,901 5,839,708, 5,864,924, 5,794,312, 5,860,191. Another method shown is the use of a pin attachment, such as in U.S. Pat. No. 4,458,384. This solution is undesirable because it damages the fabric to which it is attached. Adhesive attachment is not satisfactory because the adhesive always leaves a trace that attracts and accumulates dirt.

The present development overcomes all of these objections and provides the utility with an ease of use, inexpensive, attractive and definitely useful solution. To install a holder, one merely places the magnetic element on top of the garment and juxtaposes a keeper underneath that will instantly clamp the holder to the fabric, at which time, the temple of the glasses is inserted in the holding area provided. There is no critical alignment, the attraction is instantaneous and the removal is similarly instantaneous by simply tilting the holder forward while holding the keeper, thus breaking the field of force. The present disclosure takes advantage of a relatively new material called Rare Earth Magnets or Neodymium (NEFEM). This new material has a strength 100 times greater than the steel magnets of last century, while its size is very dramatically reduced. The field of force is so concentrated and powerful that a $\frac{1}{10}^{th}$-oz. disc magnet of Neodymium can hold more than 30 times its weight. Capitalizing on these advantages allows a very small and light holding system for eyewear to be removably attached to the outside of apparel, such as sweaters, coats, aprons, etc., in a very simple, quick method. The reversal of this simple installation procedure removes the holder and makes it available for use thousands of times.

Advantages and Objects

A method for creating a simple to operate, inexpensive to make, non-invasive, removable and reusable eyeglass holder for use on apparel, is shown and described. There are no

2

cords or chains attached to the eyeglasses, which are in the way and uncomfortable when the glasses are worn over the ears. The glasses do not jostle when walking or moving. There are no holes to be made in fine fabrics with the use of pins. There is no dependency on the availability of a pocket or button. When there is a pocket available, hanging a mechanical holder will distort or even damage said pocket. When the need is for two types of eyeglasses, one to wear and one to store temporarily, this system is very functional. Prior art shows all these disadvantages that are eliminated. By using state of the art materials, it is possible, through this invention, to avoid all these negatives. It becomes practical with this method to create a decorative, as well as functional, solution to this universal need.

BRIEF DESCRIPTION OF THE SEVERAL VIEWS OF THE DRAWINGS

FIG. **1** is a section drawing of the preferred embodiment of the invention secured magnetically to a garment, as seen from the top.

FIG. **2** is a perspective showing of the preferred embodiment showing the tapered edges, for easy eyeglass installation.

FIG. **3** is an elevation view of the preferred embodiment showing the magnets from the bottom with an eyeglass temple piece showing.

FIG. **4** is a plan view of the basic keeper plate preferably made of Alnico.

FIG. **5** is a section drawing of the basic keeper plate shown in FIG. **4**.

FIG. **6** is an alternate type of keeper plate with two magnets attached.

FIG. **7** is a section drawing of the alternate keeper plate shown in FIG. **6** as seen from side.

FIG. **8** shows a variation of FIG. **6** where the magnet carriers are movably adjustable.

FIG. **9** shows a section of drawing of another embodiment where the loose chain becomes the eyeglass holder, shown attached to a garment, as seen from the top.

FIG. **10** shows the top plan view of the embodiment shown in FIG. **9** with chain extended and more decorative detail.

FIG. **11** shows the bottom plan view of a magnet holder as shown in FIG. **9** with attaching hole for the chain.

FIG. **12** shows a side section drawing of another embodiment where there is only one magnet, shown attached to a garment, as seen from the top.

FIG. **13** shows top section drawing of the embodiment shown in FIG. **12** turned 90° C.

FIG. **14** shows a front plan view of FIG. **12** with an eyeglass temple piece inserted.

FIG. **15** shows a plan view of a smaller keeper, preferably made of Alnico.

FIG. **16** shows a section drawing of the smaller keeper preferably made of Alnico.

FIG. **17** shows a bottom plan view of another embodiment of the keeper with a magnet attached.

FIG. **18** shows the section view of FIG. **17**.

LIST OF REFERENCE NUMBERS

20—Rare Earth Magnets
22A—Preferred embodiment
22B—Second embodiment

US 6,367,126 B1

| 3 | 4 |

22C—Third embodiment
24—Eyeglass temple piece
26—Garment
28—Keeper plate
30—Tapered edge
31—Alternate style backer plate with magnets
32—Magnet housing
34—Threaded magnet housing
36—Threaded magnet receiver
38—Attaching hole for chain
40—Non-magnetic chain
42—Keeper plate

DETAILED DESCRIPTION OF THE INVENTION

FIG. 1 shows a perspective view of a preferred embodiment of a removable and reusable non-invasive magnetic force eyewear holder for use on clothing or other surfaces. The principle part No. (22A) is a holder that embodies two rare earth magnets (20) mounted on the bottom. No. (28) is a keeper plate preferably made of Alnico. When holder (22A) is placed on the outside of clothing such as a blouse, sweater, jacket, etc., and keeper No. (28) is placed underneath the apparel contiguous in location to the holder, the magnetic force between the magnets (20) and the keeper (28) will clamp the apparel and provide a holding means for eyeglasses.

OPERATION OF THE INVENTION

The operation of a preferred embodiment is, a saddle shaped unit of injection molded plastic (22A) to which two rare earth neodymium magnets (20) are affixed in wells formed to allow for insertion of the magnetic disks and the extra strong adhesive to hold them in place. This holder may be fabricated in other ways using other materials in keeping with the spirit of the embodiment. The keeper (28) in the preferred embodiment is a disk preferably made of Alnico material, that is composed of aluminum, nickel, and cobalt, together with approximately 50% of iron. The advantage of Alnico is its higher saturation capability and lower coercivity. This combination in concert with favorable weight and costing, are the advantages of its use. Keepers can also be made of ferrites or other magnetic saturable materials; however, each has its disadvantages such as the brittleness of ferrite, cost, weight of other materials.

Other species of this invention are shown in (FIG. 9) & (FIG. 12). In FIG. (9) a holder is created with two pieces, each of which has a rare earth magnet enclosed and the two pieces are attached with a chain of non-magnetic material. These parts can be molded or cast or stamped or of semiprecious stones, etc. and will function in the same manner as the preferred embodiment. The advantage of this specie is that it can be made more attractive and decorative by using precious metal or stones. Keepers can be formed with rare earth magnets attached, that may or may not be adjustable (see FIG. 8) & (FIG. 17). The advantage of the keeper with magnets is a multiplication of the force pattern and, therefore, the greater holding strength for heavier apparel or security of the more expensive holder parts. Another species shown in FIG. (12) shows a holder with one rare earth magnet and a formed opening into which eyeglass temple pieces can be inserted. The keeper piece is an appropriately sized, preferably Alnico part, that may also have a rare earth magnet (see FIG. 17) attached for applications where the holder would encounter rough usage.

Attachment of the holder (22C) is effected by bringing it into proximity of the keeper (28) placed underneath the garment. Natural magnetic attraction simplifies the procedure. No skill or training is needed. Removal of the holder requires tilting the holder downward until magnetic force pattern is broken.

CONCLUSIONS, RAMIFICATIONS AND SCOPE OF INVENTION

The reader will see that the removable and reusable non-invasive magnetic force eyewear holder for use on clothing or other surfaces is a highly reliable, lightweight, economical device that can be used by persons of any age. There is no training necessary or special agility required.

While the descriptions contain many specificities, these should not be construed as limitations on the scope of the invention, but rather an exemplification of preferred embodiments thereof. Many other variations are possible and practical to those skilled in the art in keeping with the spirit of this invention. Accordingly, the scope of the invention should be determined not by the illustrated embodiments, but by the appended claims and their legal equivalents. It should be clear that changes in size, smaller or larger, parts made of different shapes and colors and materials can be made without leaving the spirit of the invention. Likewise, it is possible to reconfigure the holding area to satisfy additional end uses. The use of the magnets in the underside piece instead of the outer holder part is also envisioned for certain applications.

The preferred embodiment as shown in FIG. 1 is comprised of a magnet carrier (22A) having a bridge (29) formed to create an open holding area (27) in which glasses may be installed, in association with a keeper (28), preferred made of Alnico. The keeper may also be provided with one or more magnets (20) either fixed or adjustable for greater holding power. (FIG. 6) (FIG. 8) & (FIG. 17).

The holder (FIG. 9) may also be made using two holders connected with a chain (FIG. 22B) instead of a single unit (22A). This format lends itself as a fashion jewelry accent since the magnets and keeper are hidden. Holder can be made of precious metal, precious stones, etc., providing a beautiful, functional holder.

The holder is also shown using a single magnet (FIG. 12) & (22C), and a smaller keeper for applications like aprons, autos, etc., or anywhere minimum size is desirable. A more powerful keeper with a magnet attached (FIG. 17) may also be used where more secure anchorage is desirable.

Another extremely valuable attribute of this system, is that the magnetic holder can be attached to almost any metal surface providing a convenient spare pair, at a file cabinet, computer area, desk area or kitchen area, thereby eliminating the need for a keeper.

The rare earth magnets are preferably neodymium (NEFEM). This relatively new material that was developed in 1983, represents the ultimate in magnetization saturation for size and cost. The great magnetic force field that the magnet can produce, therefore, controls how much it can lift or hold. Neomagnets are 10 times more powerful than ferrite magnets and more than 100 times more powerful than the steel magnets of last century. A steel magnet, weighing 16 lbs., may have a 4800 gauss and will lift 250 lbs. of iron. A neodymium disk magnet of approximately 0.500×0.125 and weighing $\frac{1}{10}^{th}$ of one ounce, can generate 2400 gauss and lift 3 lbs The other very valuable attribute of the neodymium magnet, is its very high coercivity, which makes it extremely difficult to de-magnetize. Our preferred embodiment magnets are Nickel-plated to avoid any atmospheric deterioration. Most neodymium magnets are made in Germany, Japan, China and Ireland.

US 6,367,126 B1

5

The use of this system does not create any hazard of damage to apparel no matter how delicate or coarse, nor other metal surfaces to which it is attached, and no marking will occur on clothing as in the use of adhesives.

It is anticipated that a double-sided pressure sensitive adhesive foam pad will be provided to attach the keeper to a wood or non-magnetic surface.

It is also anticipated that variations in the size and shape of holding area, together with a foam (25) or other flexible material, may be attached inside the holding area (27) for additional universality of fit for eyeglasses or other small objects weighing less than three pounds.

I claim:

1. A magnetic force eyeglass holder for releasable attachment to a garment, comprising:

two spaced-apart rare earth magnets;

a magnet housing for said rare earth magnets, said magnet housing fixing said magnets in their respective positions, said housing including a bridge interposed between said magnets and their respective positions, said housing including a bridge interposed between said magnets and forming an opening for inserting an eyeglass temple piece; and

a magnetically saturable keeper plate.

6

2. The apparatus of claim 1, wherein said at least one rare earth magnet is made of neodymium.

3. The apparatus of claim 1, wherein said at least one rare earth magnet is nickel coated.

4. The apparatus of claim 1, wherein said magnetically saturable keeper plate is fabricated from Alnico.

5. The apparatus of claim 1, wherein the opening formed by said bridge is filled with deformable foam for securely holding eyeglasses when said apparatus is jostled.

6. A magnetic force eyeglass holder for releasable attachment to a garment, comprising:

two spaced-apart magnets;

a magnet housing for said magnets, said magnet housing fixing said magnets in their respective positions, said housing including a bridge interposed between said magnets and forming an opening for inserting an eyeglass temple piece; and

a magnetically saturable keeper plate.

7. The apparatus of claim 6, wherein the opening formed by said bridge is filled with deformable foam for securely holding eyeglasses when said apparatus is jostled.

* * * * *

14 Patent Search and Conclusions before Drafting a Patent Application

Prior Art

The term "prior art" is used by the US Patents and Trademarks Office (USPTO) to describe all relevant information in the public domain on a given patentable idea or product. USPTO examiners of applications for patents would do a rigorous search of the prior art to ensure that the Claims in the application for a patent is *not* found in prior art. Information already found or disclosed in prior art cannot be patented. To be granted a patent, the application must contain material that is above and beyond prior art.

Prior art consists of all issued patents (both currently active and expired), abandoned patent applications, live patent applications, publications in books, magazines, or other media, common practice, common knowledge, and other such materials. Before preparing a patent application, it is wise to conduct a prior art search to see if the invention has a reasonable case for patenting.

Prior art search always includes a patent search using established patent databases such as Google Patents, etc. as well as the USPTO website or in USPTO designated libraries, including university libraries.

Access to Patents Database

A diligent patent search is a must before preparing a patent application. Today, one could use the database of the USPTO on the Web. US universities designated for access to USPTO databases have librarians with skill and expertise to assist an inventor attempting a patent search at the library. A new inventor is encouraged to make use of this resource in a university campus near the inventor.

Additionally, Google Patents gives easy and quick access to a database of all granted patents as well as patent applications. Other private parties also have made patents accessible to inventors via the Internet.

Search Procedures

Patent search could be conducted on a US patents database using keywords, or numeric USPTO classifications and subclassifications. For US patent 8,733,093, the keywords can be "hydrokinetic energy converters," "water energy converters," "water turbines," and so on.

To describe the numeric classification system of the USPTO, let us use the granted patent US 8,733,093. Near the bottom left side of the cover page of the granted patent you will read "USPC 60/562; 60/507; 290/42, 53." This means the granted patent conforms to US Patent Classification 60 and subclassifications 495 and 507, as well as Classification 290 and subclassifications 42 and 53. For this patent, International Classification numbers (see Int. Cl in the document) are F03B 13/18, F03B, and two others. The reader must have noticed that the US classification numbering system is not the same as the International classification numbering system, but both are printed on the cover page of all US patents.

Conclusions from the Results of a Patent Search

After a thorough patent search, the conclusions based on the findings of a patent search may fall into three categories:

1. Too many granted patents, or patent applications, or expired patents on the subject; a patent application is not recommended.
2. Moderate number of granted patents, or patent application, or expired patents on the subject; a patent application of limited scope may be justified if the patent application is prepared carefully.
3. Negligible or no patents or applications on the subject; a patent application is strongly recommended.

15 Pro Se US Patent Applications Do Succeed

This chapter is based on the author's actual experience of successfully applying, prosecuting, and obtaining several US patents as a pro se inventor-applicant. Specifically, this chapter describes the process of applying and prosecuting US Patent 7,625,152 (granted December 2009).[1] A more detailed description of the process of obtaining this patent is contained in an article published in the *John Marshall Review of Intellectual Property Law* in 2010.[2]

If one can afford a patent attorney's services to obtain a patent, it is better to do so. However, if you are an inventor who cannot afford a patent attorney's services, these are some of the choices:

1. Abandon the idea for the product and do nothing (you may be walking away from a potentially good innovation and income).
2. Start a business with your unpatented product; the risk is that a competitor may imitate your product and take away your market because of superior resources for product development, marketing, sales, and distribution.
3. Try to sell your unpatented idea to a company using a non-disclosure agreement to protect your idea (this is rarely successful).
4. Try to patent it as a pro se at a cost you can afford (one can save several thousand dollars by applying for a patent as a pro se). This would require that you study issued patents carefully and pattern your application accordingly; there is a lot you can learn from the millions of patents issued by the USPTO. The risk you face is that, in some of your applications, you may fail to patent all deserved claims (i.e., all aspects of the idea) that you could have patented, or someone

---

[1] Swamidass, P., "Does the US patent office snub inventors who file without an attorney?" InventorsDigest.com, April 2010, http://idmagazine.wpengine.com/articles/does-the-patent-office-snub-inventors-who-file-without-an-attorney/ and "Prefabricated plastic raised rumble strips and edge line for roadways," United States Patent 7625152, granted December 2009, inventor Paul Swamidass.

[2] This chapter is based on a larger, more detailed article about pro se application and prosecution of the patent application; "Reforming the USPTO to comply with MPEP section 707.07(J) to give a fair shake to pro se inventor-applicants," *John Marshall Review of Intellectual Property Law*, 9(4), 2010, 880–911.

intent on working around your patent may find a way around it; the risk is small and may be worth taking if the options 1–2 on this list are unattractive. This is a viable option for thousands of individual pro se applicants in the United States, including the author.

Seek the Help of the Examiner under MPEP 707.07(j)

Many potential inventors may be unaware that the USPTO's Manual of Patent Examining Procedure (MPEP) is friendly to pro se applicants. It requires the following seven actions from examiners of pro se applications:

1. Draft one or more claims when justified.
2. Indicate such claims will be allowed if incorporated.
3. Do not stop with bare rejection of claims; office actions should be constructive.
4. Offer definite suggestions for correction.
5. Indicate the desirability of an interview.
6. Note in office actions that certain features have not been claimed.
7. State that dependent or canceled claims will be allowed if written as independent claims.

The author found that, for the three Nonprovisional Utility patents granted to him during the 2009–2014 period, he asked for and received help with rewriting the claims disallowed by the examiners under MPEP 707.07(j). The entire process was a cordial experience with the examiners over the phone and e-mail exchanges, and one visit. It has encouraged the author to continue to apply as a pro se applicant; several applications are pending and more applications are in the process in 2016. Conclusion: Yes, MPEP 707.07(j) works! With experience, it can be faster than working with busy patent attorneys.

Several of my students taking the Engineer-Entrepreneur course using the contents of this book, apply for Provisional patents before the semester is over, or promptly after the semester. A few serial inventors attempt to write Nonprovisional applications. The new pro bono program of the USPTO enabled one students to file a Nonprovisional Utility patent through an established patent law firm in late 2015 without a fee because his low income qualified him for the pro bono program of the USPTO. Therefore, if your income level is low, visit a patent law firm in your area to investigate if you would qualify for their pro bono program to file a Utility patent application with their assistance without their usual fees. All inventive students should qualify for the pro boo program. Students, there is no excuse for not giving life to your inventive genius!

Do Not Abandon Applications Prematurely

My experience with the USPTO during the first nonprovisional application brought to the surface some issues about the examination process that may cause the pro se to abandon an application prematurely:

1. When rejections are accompanied by inappropriate references.
2. When claims are rejected repeatedly without addressing potentially allowable claims.
3. When examinations seem erratic.
4. When the examination does not conform to the spirit of MPEP 707.07(j).

My recommendation to pro se applicants is to study MPEP 707.07(j) carefully, read office actions from the examiner carefully, study relevant patents carefully, and make your case with the examiner as logically as possible until you are granted a patent; logical analysis is your ally.

The Benefits of Being a Pro Se Applicant

A pro se applicant who goes through the application and prosecution process of the USPTO to get an issued patent would find it a valuable educational experience that teaches how to write better patent applications in the future and how to prosecute and navigate the application through the USPTO office; this is invaluable. There is also a significant cost advantage if an inventor files the application as a pro se applicant. These are two reasons why a pro se inventor-applicant may be motivated to invent and patent more and more; I have seen the birth of many young serial inventors among my students, who have invented two or three patentable inventions within months after completing a one-semester course using the contents of this book. This is good for the inventor, future users of the inventor's inventions, and the economy.

PHASE 3
THE BUSINESS MODEL AND BUSINESS PLAN

CUSTOMERS, TARGET MARKETS,
COMPETITORS, AND MARKETING

16 Macroeconomics for Innovators in Engineering and Science

Macroeconomics refers to aggregated economic activity at the national level. Microeconomics refers to economic activity at a consumer or company level. It is important to remember that what happens to the economy at the national level is likely to influence the activities of most individual businesses or companies. The most notable feature of macroeconomic activity is the fact that it goes up and down in a cycle of unpredictable magnitude and duration. In other words, the macro economy may grow some years and may shrink (also called recession) during some years.

The size of the economy is measured by Gross Domestic Product (GDP), and it is used for projections of demand and sales for individual firms, as well as for revenue projections for one or more years. Thus, year-to-year GDP growth rate is an important input for business planning and forecasting conducted by private businesses. If GDP is projected to grow at a two percent rate in the next 12 months, and if your business usually grows, year to year, at the same rate as the GDP, then your projection for sales growth must take into account the GDP projected growth rate for the coming year, unless your company takes tangible steps to increase its sales through new marketing programs, if it wants to project sales growth that is larger than GDP growth rate for the coming year.

GDP, in simplified terms, is the nation's sum total of all the transactions for goods and services during the year. The nations with the largest GDPs in 2011, and selected nations with smaller GDPs, are shown in Table 16.1. Nations with bigger GDPs have more trade and business activities than do nations with smaller GDPs.

Businesses Are Affected by Economic Activity

During economic growth, a business's sales are expected to grow as a result of increased demand for its products or services. To accommodate this growth may require investment in equipment, facilities, and workforce.

The opposite may be true during recession when demand and sales may shrink. Businesses deal with expected sales shrinkage by increasing advertising/marketing efforts to attract new customers; they also prepare for reduced revenues by cutting

Table 16.1. *A comparison of selected national GDPs*

2011 World GDP		$69,659 Billion (B)
EU (European Union)		$17,577 B
Rank 1	United States	$15,094 B
Rank 2	China	$7,298 B
Rank 3	Japan	$5,869 B
Rank 4	Germany	$3,577 B
Rank 5	France	$2,776 B
...		
Rank 50	Kazakhstan	$178 B
Rank 100	Estonia	$22 B
Rank 150	Swaziland	$4 B

Table 16.2. *US economic activity in 2011*

2011 US GDP	$15.32 Trillion (or $15,320 Billion)
Personal consumption expenses	$10.87 T
Goods	$3.71 T
Services	$7.2 T
Gross private domestic investment	$2.01 T
Exports	$2,113 T
Imports	$2,697 T
Net export of goods and services	- $584 B (Imports exceeded exports)
Government consumption expenses[a]	$3,021 T
Federal consumption	$1,225 T
State/local	$1,796 T

[a] Excludes Social Security, Medicare, and other benefit payments to the population.

down inventories, orders to suppliers, and workforce size; as a last resort, getting rid of equipment and facilities is also possible.

Thus, the economic activity of the nation or the region has serious consequences for a business. Planning for the future enables businesses to prepare for increases or decreases in demand due to changes in GDP forecast.

To summarize, if the macro economy goes down during a recession, it can bring down the demand and sales for a business, and the opposite is true when demand for products increase with the overall improvement in the economic activity.

Table 16.2 shows the estimates of GDP-related data for the United States at the end of 2011, based on the data from the Bureau for Economic Analysis at the US Department of Commerce.

Types of Economic Systems

The United States is a free-market capitalist system and has served as a model for other nations. This system, which dates back to the 1700s, has served the nation very well by allowing businesses to flourish and provide goods and services to the public for less and in abundance; we rarely, if ever, see shortages and long lines for products and services. Not all economic systems are alike – far from it. For example, in the former Soviet Union, with its socialist, redistributive, centrally planned economy, it was common occurrence, before 1989, for its citizens to experience shortages of even such basic necessities as bread, salt, soap, and dairy products.

Socialist and communist economic systems that existed in Soviet Union and China have undergone some changes since the early 1990s; for the most part they have been replaced by economic systems that are more free market-friendly and capitalistic in nature, but with various levels of government, from federal to local, exercising considerable control through restrictive regulations.

In the truly free-market capitalist economy, the government generally tries to stay out of the production and distribution of goods and services (save for those regulations necessary to ensure the safety and well-being of consumers) and does not interfere with demand, capital formation, and capital investment. This allows private capital to accumulate and search for business opportunities offered by unmet customer needs; businesses form and grow rapidly to meet customer demands with new job creation as one of the beneficial consequences. The demand for new types and varieties of products creates a need for engineers who can design and develop newer and better products.

Compared to a socialist or communist economic system, a free-market capitalist economy grows rapidly. The best example is that of Communist China, which permitted free-market economy to thrive since 1978. Allowing private enterprise in production and the elimination of tight controls over production has transformed China's economy into the second-largest economy in the world behind only the United States in 2016. From 1978 to 2005, the Chinese GDP grew at an average rate of 9.7 percent a year; by comparison, the US real annual GDP growth was around 3% for the same period, and that of the United Kingdom was about 2.5%. This discrepancy is explained by the fact that both the US and UK economies are mature capitalist systems that enjoyed free-market conditions for centuries. China, in contrast, was lagging far behind, thus had more untapped potential for rapid growth that was made evident since 1978.

Here are some of the highlights of the Chinese economic transformation in the period between 1978 and 2005[1]:

1. Chinese GDP per capita increased from 2.7 percent to 15.7 percent of US GDP per capita.
2. Chinese GDP increased from 53.7 percent to 188.5 percent of Indian GDP per capita; India has not been as aggressive as China has been in opening more parts of its economy to the free market.
3. Average wages rose six fold.
4. Absolute poverty declined from 41 percent of the population to 5 percent from 1978 to 2001.

---

[1] Brandt, Loren et al., *China's Great Transformation*. Cambridge: Cambridge University Press, 2008.

In a growing economy, businesses will see demand for their products and services increase rapidly, generating plenty of profits and consequently fuelling even more growth through further investment and expansion. Start-ups, therefore, have a better chance of survival in a growing economy than in a stagnant one or, worse, in a recession. Therefore, you will see more start-ups in a growing economy.

Economic as well as employment growth occurs in a free-market economy when investment capital is allowed to form freely and generously in a climate of low taxes, low government interference, and near-term and long-term political stability. Investment capital flows into such countries from other nations, especially from the ones experiencing slower economic growth and less political stability.

In summary, small business and start up business owners must understand how macro economy influences individual business markets, demand, and profits. Businesses must learn to anticipate macroeconomic ups and downs and learn to cope or overcome them by taking every opportunity for smart growth, as well as by preparing early for economic downturns.

17   Customers, Target Markets, and Marketing

Markets refer broadly to the potential buyers for your products and services in all regions. In a broad sense, the planning and effort employed to get potential customers and former customers to become aware of your product and buy your company's products and/or services is called marketing.

Target Markets

Target market refers to those potential customers who are more likely to use your product or service; in the case of tennis racquets, for example, it would refer to tennis players or someone likely to take up tennis. Non–tennis players will be outside your target market. A tennis-racquet maker would tailor products and marketing to address the target market. Every business must seek to find its entire target market and try to reach it through marketing efforts.

(© Mindscanner / Shutterstock, 213783340)

Five P's of Marketing

Marketing efforts include communicating in a convincing manner to current and potential buyers about the company's products and services and convincing them to buy the company's products. In this process of communicating with customers, marketing may use many different channels of communication such as print or broadcasting media, direct mail, Internet, word-of-mouth, coupons, and so forth.

Markets for a company's products need not remain stagnant. A wise company is always looking for options and venues to expand the markets for its products. Expanded markets lead to growth in the company's demand, growth in revenue, and growth in profits – all marks of a healthy business.

The ultimate goal of marketing is to attract customers by convincing all potential buyers of a company's products and services to actually make purchases and enable the company to meet its sales goals. To accomplish this, marketing efforts are devoted to what is normally called the five P's of successful marketing, which are:

1. **Product:** you need products to create a market; engineers help the company design and make the best product possible for customers; make the product appeal to the target market or to enlarge the target market. A good marketing campaign must include a clear description of the product for the potential customer to understand fully.

2. **Price:** refers to an attractive price charged by the company for its products or services offered to customers; most purchases are influenced by price; if the product is sold in high-volume mass market, price needs to be at the low end among competing products; for example, McDonalds restaurants are selling high-volume fast food, and therefore the prices for its food offerings must be in the low end – that is, their products must be inexpensive from the customer's point of view; in contrast, for some products such as perfumes and designer products, matching competitors' high prices for products of similar type and/or quality is not uncommon. A marketing campaign, for most items, must give a good idea of the price of the product/service to potential buyers to enable them to make a decision to seek out your product or service.

3. **Promotion:** refers to the temporary activities of the business to bring a product or service to the attention of the target customer on a timely basis to enable the customer to make an educated or an emotional decision to purchase what the company is selling; promotion should be tied to the nature of the product – promoting a car is not the same as promoting a pack of chewing gum. An example of a short-term promotion would be 10% discount coupon, good for two weeks, distributed by a local restaurant or through the newspaper.

4. **Physical Distribution:** refers to getting the product or service to the customer on a timely basis as expected by the customer; this may require warehousing parts or finished products, shipping arrangements to get the products or services to the customer on time; in the case of tennis racquets, the distribution system may include Internet sales outlets, tennis pro shops, athletic equipment stores, and discount stores; some products require a clear return policy and a physical return system for customers displeased with the product, or if the product fails in the ownership of the customers.

5. **People:** refers to marketing and salespeople who build a relationship with all present and future customers; they provide pleasing customer service so that customers will be more likely to come back to make repeat purchases; they also establish two-way communication with customers.

18 The Power of Social Media Marketing

Haitham A. Eletrabi

Over the past ten years, social media has become a significant part of our lives; recent studies showed that young adults spend on average 3.8 hours a day on social media websites (Bilton, 2014); this is based on the use of Internet, smartphones, or both. This growing collection of media is becoming a powerful tool for taking the message about new products and businesses to users of the emerging media.

The biggest, most widely known names in social media are Facebook, Twitter, Instagram, but there are also numerous others; none of them were available twenty years ago. Their proliferation and widespread use have given rise to social media marketing, which is growing at a rapid rate. There are many different social media options available today based on "likes," "hits," "tweets," and "retweets," and other methods of dissemination. Therefore, before you can choose to use social media marketing, you need to understand them.

Why Social Media Marketing Must Be Considered?

Apple Inc. hired Musa Tariq (A social media guru) in 2014 to handle its social media presence after ignoring this segment for a long time (Oster, 2014). This shows that even the most valuable company in the world then (around $700 billion) realized the potential of social media marketing (Fletcher, 2014).

Furthermore, recent research has shown that 49 percent of small businesses have found social media marketing to be their most effective marketing tool (Mendelsohn, 2012). In addition, 63% of customers prefer businesses with information that can be easily accessed on social media websites (Sophia, 2013).

In some cases, having a weak or no social media presence might hurt a business. However, social media marketing skills are not at the top of the skill list of most small business owners. In a recent study, based on their ranking, 53% of small business owners needed most help with social media marketing (Mendelsohn, 2012).

Haitham A. Eletrabi, PhD, MBA, is an inventor and an entrepreneur.

Know Your Customer

Knowing your customers helps you select your social media platform. Consider the following popular social media platforms (as of early 2015):

- **Facebook:** The most popular social media platform with more than 1 billion monthly active users.
  - ○ **Pros:** Huge coverage, all age groups, all demographics
  - ○ **Cons:** You need good filters to reach the target audience
- **Twitter:** The second-most popular platform with more than 200 million users. Twitter is more effective with the younger audiences and heavy smartphone users.
  - ○ **Pros:** Extensive coverage, real-time capability
  - ○ **Cons:** Younger demographics, limited message space
- **Instagram:** This is growing very fast and is becoming a real competitor to Twitter. Instagram is based on pictures rather than text messages, often called "Twitter for pictures." Marketing on Instagram is effective with younger audiences, women, and for businesses that can deliver their message through pictures. For example, fashion merchants, clothing merchants, photographers, and travel agencies use Instagram effectively.
  - ○ **Pros:** Power of the visual graphic, visual appeal, instantaneous
  - ○ **Cons:** Not very text-friendly
- **Google+:** This could be viewed as an alternative to Facebook.
  - ○ **Pros:** Integration with Google services like Adwords
  - ○ **Cons:** Smaller audience (as of 2015)
- **Pinterest:** Pinterest is focused on the discovery of new products and businesses. It is a great tool to drive traffic to your site and improve your search engine optimization (SEO). Its audience is tilted toward women and foodies.
  - ○ **Pros:** Great for referral, favorite among women
  - ○ **Cons:** Not friendly for nonvisual content, smaller male audience in 2015
- **LinkedIn:** It caters to professionals looking to build professional networks. LinkedIn is a great marketing medium for consultants, service providers, networkers, professionals, and cloud computing companies.
  - ○ **Pros:** Professional nature of users, in-depth, extensive content
  - ○ **Cons:** Closed network
- **Snapchat:** This app for smartphones is pretty new (it became available in 2015) but has been gaining a lot of traction. It is a medium to send pictures to someone, and the pictures self-destruct after the receiver sees them. If the receiver tries to save the picture, the sender is notified and can use Snapchat to ensure images will not be saved somewhere, forever. This app is popular with college students and young adults.
  - ○ **Pros:** Unlimited growth potential
  - ○ **Cons:** Younger demographics, time limit for the content, works with smartphones only (2015)

There are other social media platforms emerging every day for the Internet, smartphones, or both. To use social media effectively for marketing you must match your target customer with the media that best appeals to your customer.

Cost of Social Media

Most of the social media platforms charge the business owner employing the cost-per-click (CPC) model. CPC is almost the same across the different platforms (as of 2015). In other words, the price should not influence your social media platform choice; select the social media that best suits your target audience. In early 2015, the range of CPC was between $1 and $3.50 (Null, 2013).

For example, assume a company sells affordable fashion accessories for college students. A platform such as LinkedIn or Twitter may get a very low return for dollars spent. But Facebook and Instagram may be more appropriate platforms; the choice of the media should *not* be dictated by CPC.

Set Your Social Media Goals

Make sure you understand what you want from your social media presence. Some media are good at one thing but not at another. The most common goals of social media marketing are:

- Number of Visits: Number of individuals who arrive at your website
- Engagement: Interaction of your customers with the company/brand
- Follower Growth: Number of followers on your Twitter/Facebook page
- Brand Mentions: What people are saying about your brand
  - **Number of visits**: if you are looking to increase only the hits on your website, then you may need a good event to draw potential customers to it; for example, a new article in a prominent news outlet may help drive traffic to your website.
  - **Engagement**: refers to engaging your customer through a reply, comment, or retweet (depending on the platform). To engage the customers, there must be a reason to stimulate your viewers to respond. You need to be active by posting/tweeting fresh and exciting content frequently to get your audience to notice and keep interested.
  - **Brand mentions**: It refers to your brand being talked about in a positive or negative manner. There are a few tools available to monitor your brand. For example, Social Mention is a site that monitors the social media buzz around a brand or company.

An Example of Cost and Effectiveness

An actual Facebook marketing campaign to attract traffic to a crowdfunding campaign is described here. First, the customers were defined as environmentally conscious individuals and artistic people. Marketing through the Facebook was chosen because it was a good fit for the business and customers (wide audience, combination of text and images). Additionally, the entrepreneur already had a decent Facebook following.

A simple advertisement (Ad) for the campaign for EEKONOMY (the name of the start-up company) was launched. The filters in the Facebook Ad targeted only potential customers as defined above. The goal for the marketing campaign was to increase page views and drive traffic to the crowdfunding campaign Web page, which enabled viewers to make instant contributions to the campaign. Social media tools make it possible to measure the results of your social media marketing campaign. It helps the entrepreneur figure out what worked and what did not; it can help you refine your message and select the most effective media.

The following table shows the number of hits (impressions), clicks, and cost per click. The cost for the Ad was $20 (over a 24-hour span), and the Ad yielded 7,511 hits, 63 unique clicks (means different users), and about $180 in contributions over the period the Ad was running. In other words, the Return On Investment (ROI) for this campaign was 900% (180/20 x 100), making it an almost instantaneous success. The following table summarizes the result of the social media placement of this Ad in desktop computers and mobile devices and their individual effectiveness and cost per unique click; news feed on Mobile Devices was the most effective in this campaign at a reasonable cost per click.

Placement	Impressions	Clicks	Unique Clicks	Expense ($)	Cost per Unique Click ($)
Total	7,511	65	63	20	0.31746
News Feed on Desktop Computers	269	2	2	0.98	0.49
News Feed on Mobile Devices	6,187	61	59	18.53	0.314068
Right Column Ads on Desktop Computers	124	0	0	0.04	0
Right Column Ads on Home Page for Desktop Computers	932	2	2	0.04	0.22

19   Market Analysis Resources

The Importance of Market Analysis for Your Product

How big is your business opportunity? It is as big as your market. To ascertain the size of your market, you need to conduct market research, or market size estimation. Since businesses attempt to satisfy their markets, this is an important part of the start-up business development effort; if there is insufficient market, the business has no future. Every business's potential market is the collection of actual and potential buyers everywhere. A business's market can be defined by:

- Target customers
- Relevant customers
- Share of the total market

(© Dizain / Shutterstock, 285751115)

Let us assume your new business would sell pens of various sizes and shapes. Since it can be used by all people living in the United States, it has a potential customer base of about 320 million in the United States alone (2015). Given this, the size of your market can be estimated or defined in three different ways. These are:

- **Total Addressable Market** (TAM): This refers to the total revenue opportunity available, without competition or other restrictions. So, for the above company planning to start in the city of Atlanta, estimated total number of potential US customers would be 320 million.

- **Served Addressable Market** (SAM): This is a smaller subset of the TAM that may take into account five- to ten-year growth plan after the company establishes itself. For example, during the first five years of the business plan, the company may have considered saturating southeastern United States before going national. The potential market in the Southeast would be about 88 million – a fraction of the 320 million TAM.

- **Target Market**: This is a smaller subset of TAM and SAM. This refers to the revenue opportunity carefully selected by the company to focus its 100 percent effort initially to make the business successful given the cost structure, price, and other factors. For example, this company may realize that the most direct path to success lies in the revenue opportunity offered by college students in three southern states in the United States. Furthermore, the target market for the first two years may be limited to a total of, say, fourteen college campuses in the three states with a total enrollment of 400,000 students. This then is the target market until the company has the resources to market to SAM or beyond SAM. For a start-up, reaching 400,000 college students is more practical than reaching 320 million US residents. In the first two months of the start-up, however, the target market could be limited to perhaps two colleges with an enrolment of 40,000 students; a very manageable target market. Remember that the marketing budget grows with the target market. For example, if it takes $0.2 to reach each student with your marketing message, to reach 40,000 students the necessary marketing budget would be $8,000. Therefore, the available budget in the early days of the start-up would determine the practical and affordable target market.

With each business and its product, the size and nature of the target market changes. To do a reliable and thorough market research, digital and other databases are needed. Here are some data resources, most of which are digital.

1. **US government data resources**

   - Search for the North American Industry Classification System (NAICS) number for your product on the US government databases and then find the size of your industry in the US Census Bureau database. NAICS has a number for every industry (or product), which makes it easier to find the total size of market by NAICS classification for the product or business: www.census.gov/eos/www/naics/

   - US Department of Commerce, US Census Bureau has a large database; www.census.gov/#

2. **Selected market research databases at a university library (you can find them in public libraries too)**

   Bizminer
   Factiva

First Research
Passport Database (International consumer markets)
Standard and Poor's NetAdvantage
Statista – The statistics portal for market data, market research, etc.
IBISWorld – Leading publisher of industry research and procurement research
Hoovers.com – For industry analysis, sales leads, etc.
UN Comtrade
US Census Bureau: USA Trade Online

3. **Report Linker – for reports on industry**

4. **Additional Data Sources**

CIA World Factbook
NationMaster
Wikipedia
Data.gov
US Census Bureau
Barrons: www.barrons.com
The Economist: www.economist.com
Wall Street Journal: www.wsj.com

20 Illustrative Case: Market Analysis for Tennis Racquets

Market research/analysis is the process of setting a research goal, gathering data, analyzing it to understand the total market and target market, as well as drawing conclusions from the analyses to set targets/goals as part of a business model, strategic plan, or marketing plan. Data gathered for market research could be either primary or secondary; primary data is new data you collect, and secondary data is already published data accessible to the researcher for free or for a fee. The data that *you collected* from a market or consumer survey described in Chapters 8 and 9 is primary data. But it is impossible to collect all marketing data on your own; we must depend on published secondary data to complete our market research.

Secondary Data for Market Research

Good market research is data-intensive and is a function of the quality of the data, quality of analyses and interpretation of the findings. Sources of secondary market research data are industry specific. For example, if the business sells medical equipment, it will find appropriate data in publications and reports pertaining to the medical equipment industry or the hospital industry. On the other hand, the data for sports equipment such as a tennis racquet, the data concerning markets, target markets, market size and annual sales in terms of dollars or number of units will be in sports-related racquet industry publications and reports.

Secondary data for market research can be found in libraries (seek the help of a business topics librarian), US government publications for US markets, publicly available Web pages (use with caution because there are spurious Web pages with questionable data), commercial databases for a fee, reference publications by Moody's, and other leading publications on companies and industries in the United States.

US government publications for market research are many; market size–related publications are issued from the US Census Bureau of the US Department of Commerce. In their publications, industries are classified in two to six digits of the North American Industrial Classifications System (NAICS). For example, NAICS 339920 refers to sporting and athletic goods manufacturing including **tennis goods manufacturing** (balls, frames, rackets, etc.). US government publications pertaining to NAICS 339920 may have useful information for market researchers investigating the size and scope of this industry. The six-digit code is constructed as shown below:

1. 339 is Miscellaneous manufacturing;
2. 3399 is Other Miscellaneous manufacturing;
3. 33992 is Sporting and Athletic Goods manufacturing;
4. 339920A156 is for tennis equipment, excluding apparel, nets, and shoes.

Sample US Census Bureau Data, 2002

The following are sample data for the NAICS classification, 339920A156:

1. Number of companies making at least $100,000 in shipments was 8.
2. Total value of shipments for 2002 was $56,701,000.

Above data are based on the 2002 Economic Census. This is only a small sample of what is available in government documents. More data can be obtained from industry associations and trade magazines.

A Sample of Industry Publications, 2008

According to *Tennis Industry News*, the following data is about the tennis industry and growth rates for the United States, published on March 1, 2008[1]; one may need more current data while conducting market research today.

1. Racket shipments up 42.1 percent since 2003 (+1.32 million units).
2. Racket shipments in 2007 up 9.5 percent over 2006 (+390,000 units).
3. Consumer purchases over the Internet have increased in all categories.
4. In 2007, frequent players increased to 5.3 million, a 15.1 percent rise since 2003.

It may not be sufficient to use only secondary data for market research. Some primary data collected firsthand by the market researcher from potential customers, as described in Chapter 9, would provide a more complete picture of the target market and its needs.

Primary Data for Market Research

Primary data for market research/analysis can be obtained in several ways: through interviews, focus groups, observations, and surveys of potential customers. Primary data collection is frequently conducted using a custom-designed survey questionnaire (see Chapters 8 and 9). Every new business must know who its customers are, what they want, and if they would purchase the company's products. The data so collected may be processed to understand the market and customer interests; the findings may be used to fine-tune the target market and improve the product, or add

---

[1] Tennis is the Number 1 Traditional Sport, *Tennis Industry News,* www.racquetsportsindustry.com/tia/2008/03/tennis_is_the_no_1_traditional.html

product features to satisfy target customers. It can be deftly combined with secondary data analyses to get a more complete picture of the market to set goals, targets and plans for the company. The reader can find useful market survey templates at Survey Monkey, The Wufoo Forms, and many other sites on the Web.

Secondary Data Sources Used in This Chapter

A. www.census.gov/econ/census07/

B. www.hoovers.com/companyreports-marketresearch/100005815-1.html

C. Racquet Sports Industry (RSI): www.racquetsportsindustry.com/tia/2007/05/know_your_market.html

D. www.racquetsportsindustry.com/tia/2012/05/2011_state_of_the_industry_rep.html

21 Competition Research

Why Competition Research?

Engineers and scientists can be easily biased toward their products and may over-estimate the long-term success of a business founded on their invention. They must recognize two kinds of uncertainties that can take away their customers, reduce their revenues, take away their profits, and force their business to shut down. The two big uncertainties are: (1) one or more new and unknown technological innovation that may suddenly surface in the market to take away the customers from existing businesses, and (2) unexpected actions by competitors that can draw away a significant number of customers from an existing business. For a new business that is about to start with a new technology or invention, the biggest source of uncertainty and threat to its revenue comes from competitors already in the market or likely to enter the market upon a new entrant's entry into the market.

Therefore, it is essential to conduct a study of competition before a new business start-up is launched. Competition could originate from existing companies with similar or somewhat similar products and from companies not in the same market. For example, when digital cameras arrived, Kodak was an established and the largest camera and chemical filmmaker in the world. With the arrival of the digital camera, chemical films were pushed aside by digital pictures, and Kodak needed to deal with competition from traditional camera makers as well as large electronics companies that jumped into the digital camera world to take advantage of their own digital technology and adapt it to the camera industry.

As a result, Kodak continued to face competition from traditional camera makers such as Nikon, Cannon, Fuji, and others. Kodak also had to contend with electronics heavyweights such as Sony, who entered the camera industry very successfully. In the end, Kodak did not fare well against the new competitive environment and was no longer the leader in the industry; its chemical film domination was wiped out by the digital substitute.

Kodak is an example of a dominant company that lost markets, customers, and revenue because of changing technology and new competitors in the marketplace.

What Competitors Might Do When Your Product Enters the Market?

Competitors respond to new products in the market with steps to drive out the new entrant. Thus, it is essential to research the list of potential competitors in the target market. Above all, competitors must be prevented from bringing imitations of your product to the market to take away your customers and your revenue; your patents prevent competitors bringing the same product or its imitation to the market.

During competition research, document all competitors already in the target market as well as those who might enter the target market. Look for competitors who are directly and immediately threatened by your product entry to the market because they are the most likely to respond immediately to your entry into their market.

Their response could include ignoring your product as "too minor" compared to their market size. Also, competitors could drop the price significantly for their product competing against your product, thereby causing your company to cut prices to retain customers, severely undercutting your profit margins. The longer this situation lasts, the higher the likelihood of your start-up failing, unless your company is prepared for it.

Competition is not always bad. If your company brings a new product that is patent protected to a market with a major dominant company, your company could be a target for future acquisition by that larger company. This could prove to be a lucrative deal for the inventors behind the new start-up company. Apple, Microsoft, and others allow small competitors to grow but then acquire them when their business gets established and has a sound product and customer base. The acquisition by a larger firm is a desirable exit event for the inventors and investors in the new company.

Sustainable Competitive Advantage

A company needs sustainable competitive advantage to stay in business. There are several ways of attaining and maintaining sustainable competitive advantage:

1. Cost leadership – that is, low costs others cannot or would not match.
2. Differentiation – the new company's product is different in quality, technology, customization, speed of service, or some other critical factor that keeps competitors away.
3. Focus – the company is focused on being the best in something specific, and, consequently, it is difficult for others to take away their customers.

MANUFACTURING, SOURCING, DISTRIBUTION, AND REACHING THE CUSTOMER

22 Manufacturing and Sourcing

Five objectives of a manufacturing system are to attain the best in cost, quality, flexibility, delivery, and customer satisfaction (CQFDC). Goods may be manufactured or procured from a reliable subcontract manufacturer (a subcontractor or a supplier), who can ensure good quality at a cost that is profitable for the new start-up business.

The decision of new start-up business to make a product or buy from a subcontractor is called the "make or buy" decision. It should not be based on cost alone; in a recent case, some US wholesale suppliers to homebuilders in the United States imported low-cost drywall (also called, plasterboard, wallboard, or gypsum board) from another country. A few years later, the drywall was found to have a toxic substance in them and needed removal from the walls of new homes at great cost to the contractors/suppliers, and at great inconvenience to the homeowners – who did, however, receive compensation.

Figure 22.1. The steps to an excellent product

(© Asfia / Shutterstock, 256250953)

The Manufacturing Options

1. Procure finished products from a subcontract manufacturer; or
2. Procure all parts and components but do the final assembly; or
3. Make most of the parts and components, and final assembly; procurement plays a smaller role.

Traditional Manufacturing Process Options

The total volume of goods to be manufactured in a given period (example: in a given month) determines the investment and labor required to carry out manufacturing. There are traditionally three broad manufacturing possibilities:

1. Low-volume, custom manufacturing in a job shop; one unit at a time, if needed.
2. Medium-volume production in batches.
3. High-volume production (mass production).

One distinguishing feature among the three manufacturing options is the layout of the equipment and the way materials flow through the manufacturing facility. Each option is discussed in greater detail in the following sections for the benefit of the reader with no background in manufacturing.

Low-Volume, Custom Manufacturing Job Shops

In a job shop, machines or work centers are not arranged in any sequence. If you produce one or two units per customer, a job shop is appropriate. Automotive repair shop is a job shop (it is a case of repair service, yet a good example). The auto shop takes the first customer's order and fixes that which is wrong in his/her car, maybe a 2008 Ford Explorer, which needs the disc brakes replaced. The next person may bring to the same shop a 2010 Jeep Grand Cherokee for a radiator work. A job shop can handle all these unique demands with skilled employees, who can diagnose the problem correctly and fix it with general purpose, generic tools and equipment. This is an expensive production system for making a high variety of products with skilled labor, where the cost per unit is higher than all other production systems.

In manufacturing circumstances, job shops are more common for special-ordered large or complex equipment. One obvious example is the Orange County Choppers (OCC), which has a TV show describing the manufacture of unique custom motorcycles. Another example, Express Oil Change (Figure 22.2) is a company in the USA that does custom oil changes as well as auto repairs using the slogan, "10 minute service." The company can change oil in 10 minutes for cars of all models and makes arriving in any order. Their skilled employees can handle cars of any model, make and size.

Medium-Volume Batch Production

A manufacturer may produce a product or components in batches of 100 or 1,000, and so on. Some production facilities are designed to produce a batch for one customer and then switch production for the next customer with or without some variations in the product. For example, if a specialty shoemaker gets an order for one batch of 1,000 brown shoes from one customer in Texas, and another order of 600 black shoes from a customer in New York, the company may get raw materials prepared for 1,000 shoes of the Texas order (brown shoes) and devote all machines and employees of

Figure 22.2. A custom oil-change shop

(Copyright © author)

the department to making all the Texas shoes first. Once the Texas brown shoes are done, raw materials for the 600 black shoes for New York are prepared, and the shoes are manufactured by the same employees and the same machines. In medium-volume batch production, there can be variations in the batch size (1,000 for one and 600 for the other) and variations in the product size, color, shape, dimensions, and other aspects. Batch production serves subcontract manufacturers very well because they may serve as subcontract manufacturers for several customers.

High-Volume Mass Production

High-volume mass production often uses dedicated production equipment (at very high initial investment) to make one product, with minor variations. Coca-Cola bottling is an example of high-volume mass production. Millions of identical Coca-Cola cans or bottles filled with the same beverage may be produced in a single bottling plant working 24 hours a day, seven days a week. This is a highly efficient system for low-cost high-volume products. Robotic assembly lines can increase the variety and rate of production in mass production lines.

Cellular Manufacturing

This is a nontraditional production system, which is more flexible and yet more efficient than traditional systems. It can boast the benefits of both the job shop and high-volume production. Cellular production is flexible to make a variety of products in a given cell and also flexible enough to rearrange the cell if the product changes significantly. Cellular production accomplishes high-efficiency and high-variety production by reducing or eliminating setup costs, wait times, and inventories and by attaining smooth flow of products through the cell.

Cellular assemblies with machines or entirely manual operators have worked very well. Robots in cellular manufacturing have contributed to increased production rates and flexibility to change the products easily.

Contract Manufacturing

There are contract manufacturers for most products to supply to new start-up businesses. Although, a product may be made by a contract manufacturer for a new start-up, the product may carry visible identification and branding of the new business. For example, if a product is made by a contract manufacturer in a foreign nation, the product could carry the brand name of a new start-up business in the USA. Some businesses may use contract manufacturers during peak demand seasons as a way of adding capacity to their production without making investment in new facilities and without hiring new employees.

Design for Manufacturing

Design for manufacturing is a principle for product design that simplifies the product during the design stage to reduce manufacturing costs and increase product quality and ease of manufacturing. To achieve true design for manufacturing, the designer works closely with manufacturing engineers from the production facility entrusted with the manufacturing of the product.

23 Break-Even Analysis

The cost of a manufactured product is composed of fixed costs (not affected by total quantity produced during the year) and variable costs (vary directly with the quantity produced during the year). The relationship between fixed and variable costs for a product determines when the company will begin to make profits. At a break-even sales volume, a company's total costs are fully covered by sales revenue.

The underlying relationship used in computing the break-even (BE) quantity is TR = TC (that is, Total Revenue equals Total Cost at break-even quantity); this relationship can be used to compute the break-even quantity. If P is the selling price, BE is the break-even quantity, TFC is total fixed cost, and V is the variable cost per unit, then:

$$\text{Given: TR = TC} \qquad (1)$$
$$\text{Total variable cost (TVC) at BE} = V \times BE \qquad (2)$$
$$\text{At break-even quantity, BE:}$$
$$\text{Total revenue, } P \times BE = TC$$
$$= TFC + TVC$$
$$\text{Since TVC} = V \times BE$$
$$P \times BE = TFC + (V \times BE)$$
$$P \times BE - (V \times BE) = TFC$$
$$BE (P - V) = TFC$$
$$\text{Therefore, BE} = (TFC) / (P - V) \qquad (3)$$

Thus, we can compute the break-even (BE) quantity using equation (3). There are three ways of ensuring that the break-even volume is below the total sales volume to guarantee profits for the business. They are:

A) reduce the total fixed cost TFC;
B) increase the selling price P;
C) reduce the variable cost per unit V; or
D) all of the above.

An Illustration

In the example in the following table, the BE volume can be computed with the data given below:

P = $3.0
V= $1.2
TFC= $259,000
Therefore, BE = $259,000 / ($3 – $1.2) = 143,889 units

Variable cost per unit is computed by (total variable cost) / (total sales for the year) = $1,200,000/1,000,000 = $1.2. At a BE quantity of 143,889, the company is generously profitable at the current annual sales quantity of 1 million units per year.

This hypothetical business is very healthy because even if total sales quantity were to drop to 700,000 (30 percent drop in sales), the company will still make healthy profits, because the BE quantity 143,888 would remain well below the sales quantity of 700,000 units.

Given: Total sales = 1,000,000 units/yr.

Cost Description	Fixed Costs	Variable Costs
Variable Costs		
Cost of Goods Sold (excluding direct labor cost)		$1,000,000
Inventory		$0
Raw Materials		$0
Direct Labor (Includes Payroll Taxes)		$200,000
Fixed Costs		
Salaries (of managers, includes payroll taxes)	$150,000	
Supplies	$200	
Repairs & maintenance	$0	
Advertising	$30,000	
Car, delivery, and travel	$2,000	
Accounting and legal	$5,000	
Rent	$5,000	
Telephone	$1,500	
Utilities	$1,500	
Insurance	$2,000	
Taxes (Real estate, etc.)	$2,000	
Interest	$10,000	
Depreciation	$0	
Other (specify)	$0	
Miscellaneous expenses	$0	
Principal portion of debt payment	$0	
Owner's draw (salary)	$50,000	
Total Fixed Costs (TFC)	$259,200	
Total Variable Costs		$1,200,000
Variable cost/unit (V)	1.20[a]	
Selling price (P)	$3	
Break-Even units (BE)	143,889	

[a] $120,000/100,000 = $1.2.

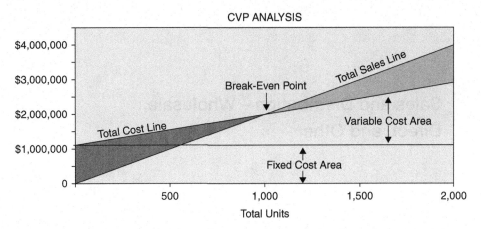

Figure 23.1. Break-Even or CVP Analysis

(Used with permission: www.principlesofaccounting.com/chapter18/chapter18.html; Dr. Larry Walther, Professor and Head of the School of Accountancy, Utah State University, UT.)

Some authors use the term "cost-volume-profit (CVP) analysis" to describe break-even analysis, which can be expressed graphically for clarity as well. Figure 23.1 shows that fixed costs are $1,000,000, as shown on the y-axis; they are fixed because they do not increase or decrease as total number units produced/sold increases from 0 to 2000 on the x-axis.

The Total Cost Line includes Fixed Cost and Variable Cost (the Variable Cost line is stacked on top of the Fixed Cost line). The Total Variable Cost Line and the Total Sales Line increase steadily as sales increases from 0 to 2000. They intersect at the Break-Even Point, which is 1,000 units. To the left of the Break-Even Point, total cost is more than total sales, resulting in losses until 1,000 units are produced and sold. When production and selling exceed 1,000 units, total sales are more than total costs, resulting in profits. Therefore, to operate profitably, a business must sell more than the BE quantity – more sales over the BE quantity, the better.

24 Sales and Distribution – Wholesale, Direct, and Other

Marketing is distinct from sales and distribution. Marketing's function is to make potential customers aware of the company's products, their functions, features and price, and where to get them. Marketing function requires a sizable budget for most businesses. Once marketing has accomplished its goal, customers wanting to see, inspect and buy your product must be addressed via sales and distribution functions.

Marketing

When you drive along an interstate freeway in the United States, you will see very tall billboards advertising one product or the other. Figure 24.1 gives an example of such a billboard near a freeway exit. This is the work of the company's marketing department. It conveys to potential customers driving in their cars the following information to encourage them to make a decision to go to the nearest McDonald's restaurant:

1. The brand (red background and golden arches).
2. The product (iced mocha and latte, McRibs, and 42 oz. drink).
3. The price (99 cents).
4. Not shown:
   a. Where (Example, Exit 28).
   b. When (Example, available 24 hours).

Figure 24.1. Billboard advertisement

(Copyright © ffooter / iStock, 36329400)

Distribution Channels and Sales

A driver who responds positively to the billboard in Figure 24.1 would take exit 28 and find the McDonald's restaurant, which is the sales outlet, where the new customer would transact with the sales staff, who would sell him his order. In this example, marketing communicates to the customer using the billboard, and sales staff complete the sale and transaction at the retail store.

Let us revisit the tennis racquet case described earlier in the book. Tennis racquets are sold through several distinct channels in the United States:

1. Internet-based stores such as, Tennis-Warehouse.com, TennisExpress.com, etc.
2. Sports equipment stores such as Sports Authority, Academy Sports, Modell's Sporting Goods, etc.
3. Tennis pro shops in tennis clubs and city recreation centers.
4. Discount stores such as Walmart.

(© Dizain / Shutterstock, 246641950)

Sports equipment stores allow players to borrow "demo" racquets for a few days and play with them before making a purchase. It is common for players to borrow more than one to compare performance of the racquets. Even Internet-based tennis equipment stores now permit players to "demo" several racquets for a week, for a deposit. At pro shops attached to large tennis facilities, the player could discuss the racquet properties with the club's tennis professional. Advanced players who know their own playing styles and racquet preferences use Internet stores to purchase racquets at a discount.

Distribution refers to the process of getting the racquets to the ultimate consumer – with or without intermediaries. In the above case, tennis racquet makers such as Prince, Head, Wilson, and others rarely sell directly to the player. Therefore, tennis racquets are being distributed through discount stores, Internet stores, sports equipment stores, large retailers, and pro shops. All these outlets may obtain their racquets from wholesalers or the manufacturer. This example of tennis racquet sales and distribution uses diversified channels to reach the tennis-playing consumer by various means.

The goal of sales and distribution functions of a company is to ensure that the customer is completely satisfied and the various distribution channel owners are profitable. Through customer satisfaction sellers try to turn all past customers into repeat customers.

Decisions Concerning Distribution and Sales for Start-Up Companies

Distribution and sales can cause customer dissatisfaction if not planned and executed well. Here is a list of decisions to consider:

1. Have a reliable system to process customer orders in a timely manner.
2. Use all relevant and efficient ways of distributing the product or service: wholesale distributors, retailers, direct sales via the Internet (Amazon or EBay are examples), and/or other methods.
3. If your company takes orders directly from customers, make sure you have well-trained staff who will field calls, take the orders, and fulfill them in a timely manner and accurately.
4. Set up a payment system that is convenient for the buyer as well as reliable and secure for your business.
5. Set up a customer service system to deal with dissatisfied customers.
6. Carefully analyze the costs and benefits of free shipping, full or partial refunds, and other measures that would allow your business to potentially expand repeat customer base but, remember, these steps cut into your profit margins by adding extra costs.

25 Reaching Your Customer: Advertising and Promotion

Reaching Your Target Market

Advertising and promotion are customer acquisition tools of the overall marketing strategy. Both incur costs. Without them, however, a new business cannot get the word out to its target customers.

There are multiple media options for reaching the target market. Each one has a different cost and different reach. Cost of reaching your customers is critical in the early days of a start-up; the available funds must be allocated among the various media wisely for maximum effect for the dollars spent. The following are major media for advertising to reach your target customers:

1. Direct mail and telephone
2. Radio
3. Television, local or national
4. The Internet
5. Newspapers (mostly local)
6. Magazines (weekly or monthly)
7. Billboards (see previous chapter)
8. Other creative channels

(© Dizain / Shutterstock, 284806601)

Promotion

While advertising has relatively long-term results, promotion is for the short-term goal of compelling established or new customers to buy your product/service; it could include free product samples or coupons with a deadline. The ultimate purpose of a promotion effort is to build sales in a short time. The cost of the promotion may cut into the profits in the short term, but the business is likely to benefit from exposure to new customers in the long term.

The Cost of Reaching a Customer

Ultimately, for a new business start-up, three issues are important. First is the careful selection of the right combination of media in which to place advertisements adequate to reach and create awareness among target customers. Second is inducing sufficient numbers of targeted customers to purchase, so revenue from sales equal or exceed your sales goals. Finally, if the first two issues are satisfied, the third consideration is ensuring that the cost of advertising and promotion is low enough to enable the business achieve the projected profits.

It is not difficult to compute the cost of reaching a customer annually; the total advertising expense for the year divided by the number of customers who actually purchased your product is the cost of advertising per customer. For example, if the advertising expenses for the year were $20,000 and actually 1,000 customers purchased your products in the year, the cost of reaching one customer was $20,000/1,000 = $20; although you may not know how many saw the advertisements.

At the beginning of the year, while budgeting, one has to make a reasonable estimate of the cost per customer and periodically reassess the response to advertisements and promotions and adjust the marketing budget or the media used to reach customers. This is a rather approximate science.

SELECT YOUR BUSINESS MODEL

26 | Selecting the Pre-Start-Up Business Model

A business model defines how the business would function smoothly with all its parts and components to satisfy customers and generate revenue and adequate profits for investors or owners of the business.

Let us use the automobile as an illustration of the business model. An automobile consists of the frame, body, seats, steering, engine, transmission, axles, wheels, tires, and many more parts. Each component of the automobile functions in relation to all other components of the car to move the driver in the car from one place to another whenever the driver wants to travel. Similarly, a business model is composed of many distinct components of the business to enable it to keep customers satisfied and generate revenue and profits as planned. Knowing all the parts of the proposed business and how they all fit together could be a challenge for engineering students and engineers. However, it not impossible for engineers to learn how a business model is put together.

(© Master_art / Shutterstock, 242671303)

The business model is the selection of key components of the business to accomplish the desired goals of the business. A carefully selected business model (B-Model) becomes the basis for a business plan (B-Plan) that follows. Often, start-ups need financing from investors; without an investor, some start-ups cannot take off. From an investor's perspective, before deciding to invest, an investor would look for a business plan, which is built on one carefully selected B-Model among many possible options.

The careful selection of the B-Model increases the probability of success of the start-up, if the start-up becomes a reality. A flawed business model would ensure the failure of the start-up. However, as the start-up takes shape, the business model may need adjustments, and the entrepreneur must be sensitive to the needed changes and be prepared to act on them. For example, the target market, originally envisaged in the B-Model may be inappropriate and may need revision based on more recent information.

The challenge to the entrepreneur is to select *one* robust B-Model among many viable options. During the pre-start-up phases described in this book, an entrepreneur must devote time to the selection, testing, and refining of the business model before completing the business plan that follows. Business model development consists of the search for answers for a number of questions and issues; most are listed below.

Once a B-Model is developed and is reasonably firm, the B-Plan would follow. You will find Larry Kim's article[1] on downloadable top-ten business plan templates quite useful before creating a business plan from your business model. The 50+ questions in the list below should prepare an engineer adequately to attempt the business model—not all questions may apply to every start-up business.

Fifty+ Questions to Ask before Creating a Business Model

(Not all questions may apply to each and every start-up)

The Product/Service

1. Are you entering a hot market? Is the product a fad?
2. What do you offer to the customer? Is it unique? Could others offer the same at a better price and quality?
3. Is there a compelling reason to buy your product/service?
4. Is this a price-sensitive product/service?
5. Is this a mass-produced or low-volume business, or in between?
6. Is this a custom-produced product?
7. Do you make the product, do you assemble it, or do you get it fully produced and delivered by a supplier?

---

[1] Kim, L., Top 10 business plan templates you can download free, *Inc.*, June 11, 2015, www.inc.com/larry-kim/top-10-business-plan-templates-you-can-download-free.html

(© DelMosz / Shutterstock,
114827764)

Customers

8. What are the customers' needs or problems that you propose to solve?
9. What is the value of your product or service to the customer?
10. Who are the customers needing this product or service?
11. What value would your customers assign to your product or service?
    11a. How much would customers pay for it?
    11b. Is this a sustainable benefit to the customer? A "pet rock" provided unsustainable value – consequently, the business rose and fell.
12. How do you find and reach the customer?
13. Are there clear prospective customers?
14. How large is the customer base?
15. Do you have access to prospective customers?
16. Is the potential market not served or underserved now?
17. Do you need a strong brand to attract customers? How expensive would it be to create a new brand identity?
18. Do you plan to use warranties to attract customers and keep them satisfied? Do you have a manageable system for handling the warranties and the resources for satisfying customers?
19. Could you create demand if none exists today?
20. How would you develop and retain customers?

Markets and Marketing

21. Is the business model scalable? Is there enough demand to make this a full-time business to pay employees wage and produce a return on investment of at least 35 percent a year?
22. Do you have a plan to introduce the product and operate in a given geographical region?

23. Is this product/service meant for a large geographical area – city, state, region, nation, or international? Do you have a budget to reach customers scattered over a large area?

24. Is your business restricted to a small area – say, a college campus? How would you grow if you needed to increase your revenue?

25. How many and which marketing channels do you propose to use?

26. What would be the qualifications of sales personnel, and how would you train them?

Manufacturing or Sourcing, Distribution, and Sales

27. If you start by procuring a fully assembled product from a supplier, is the supplier capable of scaling up if you find the market growing rapidly? If not, what would you do?

28. How would the product go from manufacturing to the customer?

29. What are the costs involved in moving the product from manufacturing to customers?

30. How would the customer purchase and pay for the product?

31. Where would the customer purchase the product?

Competition

32. How formidable is the current competition?

33. Could you keep competitors away from your business? If yes, for how long?

34. Do you expect competition to engage in unfair practices to keep you out of the market?

35. If you operate in a niche, how long could you protect the niche from competitors?

36. What do you expect your gross margin to be? Do you foresee an advantage over your competitors? (gross margin is 100 x (revenue minus cost-of-goods-sold)/ revenue; it is a percentage of the total revenue left over after all direct costs are deducted from revenue; the leftover amount could be then used for selling expenses, general administration, research & development, interest payments, and dividend distributions to investors; the larger the margin, the more attractive the start-up business would be to investors).

37. How good is your intellectual property protection to keep competitors away?

38. Could you gather reliable data about current and potential competitors?

Finance and Profitability

39. Is the break-even sufficiently low compared to the size of the market and competition?

40. What is the cost of reaching each customer?

41. How is value added step by step in the value chain?

42. Do you have the skills to handle the financial side of the start-up, or do you have trusted people to do it for you?
43. Do you have the resources for the pre-start-up phase?

Staffing and Management

44. Can you attract talented employees and business partners?
45. Do you have a skilled and trusted team?
46. Do you have the technical people to develop a high-quality products and keep the products ahead of competitors, and take on challenges from competitors?

Uncertainties

47. How well could you predict your business activity (demand, cost, etc.) for the next five years?
48. What do you expect your business valuation to be in five years? Would you expect your business to be valued at $1-5 million five years from now? If not, what do you expect it to be? What do you want it to be?
49. Are there cumbersome government regulations (for example, FDA or EPA) to overcome before market entry, or after market entry? Do you have a plan to deal with them? Do you have the expertise and resources?
50. Could you face legal obstacles to this business? How would you mitigate legal issues?
51. Can you handle losses? How big a loss could you handle without going under and totally destabilizing yourself financially? What would be a fatal loss?

Strategic Issues

52. How would the start-up use its resources against competitors for sustained success?
53. What are the core competencies of the start-up?
    A. Is it from the intellectual property and/or proprietary trade secrets?
    B. Is it from the skills of the staff/management/leadership/technical personnel/network?
    C. Is it from the sourcing and distribution system?
    D. Is it from strategic partners?
    E. Is it from first-mover advantage?
    F. Is it from the efficiency of production and distribution?
    G. Is it from cost or pricing opportunities?
    H. Is it from unique company resources? What are they?
    I. Is it from unique processes?
54. Have you tested and validated assumptions, variables and relationships of the business model? Are there key business assumptions that are still untested?
55. Do you have access to reliable customers to test and validate your assumptions?

Exit from the Start-Up

56. Do you have a plan to exit the business? When? How?

How to Pick a B-model among Various Options?

Using the answers to as many questions as possible, select and assemble the components of your business model with the goal of ensuring customer satisfaction, your revenue and profits. Remember that the B-Model must assume many things about the proposed business. There is a danger that wishful thinking of the entrepreneur may become assumptions underlying the business model; that could lead to a disaster.

To prevent wishful thinking taking over the business model, test and validate your assumptions with factual data. Critical assumptions about customers, the market, costs and price must be tested and validated. For example, to test and validate the nature of customers for your product/service, conduct a customer or market survey; this book provides examples of customer survey in chapters 8 and 9.

In summary, test and validate all key business model assumptions and pick a business model that is most robust and likely to be stable under various circumstances, and has a sizeable initial market with strong growth potential.

Other Ideas on B-Model Development

Other teachers such as Osterwalder and Pigneur (2010) and Blank and Dorf (2010) have popularized business model development. Many of the answers to the questions above should give you ammunition to put together a business model regardless of the teacher you wish to follow. The reader can benefit from the teachings of multiple authors on the subject. For example, the Business Model Canvas proposed by Osterwalder and Pigneur (2010) covers the following list of decisions and their interconnections on a two-dimensional chart:

A List of Decisions for the Business Model Canvas

1. Key partners – suppliers, raw material providers, service providers
2. Key activities – used to build value for customers
3. Key resources – that add value to customers, distribution, customer service, etc.
4. Value propositions – customer problems being solved, needs satisfied, product features, etc.
5. Customer relations – getting, keeping and growing a customer base, cost of acquisition per customer, etc.
6. Customer segments – targeted, important and unique customer base
7. Marketing channels – different ways of getting information about product/service to customers to describe uniqueness, distinction from competition, value proposition, etc.; evaluate effective and/or efficient channels; see Chapter 18 for examples

8. Cost structure – important and critical costs, cost controls, cost advantage over competition
9. Revenue streams – customer desire to pay, current market price, pricing for the short and long term, price sensitivity of customers, expected stable revenues, and break-even threshold.
10. Describe the relationship between VALUE PROPOSITIONS AND CUSTOMER SEGMENTS
11. Describe the relationship among VALUE PROPOSITIONS, CUSTOMER RELATIONS and MARKETING CHANNELS
12. Describe key relationships among CUSTOMER RELATIONS, CUSTOMER SEGMENTS, MARKETING CHANNELS AND REVENUE STREAMS
13. Describe important relations among KEY PARTNERS, KEY ACTIVITIES, COST STRUCTURE AND KEY RESOURCES

ESTIMATED COST, PRICE, AND PROFITS

27 Key Decisions: Costs Estimation and Pricing

Cost Estimation

For an engineer and scientist assembling a business plan, cost estimation may be a difficult exercise, but far from impossible. The goal is not to be 100% accurate. The goal is to get a cost estimate, which is in the general region of real costs. A business plan requires multiple estimates, each estimate confirming but not disconfirming other estimates; cost estimation is a big jigsaw puzzle for a start-up business. You can find a lot of useful information about business cost estimation on the Web, though it is not the only source for such information.

(© Dizain / Shutterstock, 371912968)

Total unit costs can be inferred by working back from similar products in the market. If a similar product retails for $20, and if the estimate of the markup for such products is 30% at the retail level, one can estimate the cost of the product to be $14 to the retailer. Similarly working back, through wholesaler the cost to the manufacturer could be estimated. If the approximate cost to the manufacturer is estimated as $8 per unit, the new start-up would have an approximate target cost of $8 to match or beat, until a better estimate of the cost comes in.

If the business plan calls for sourcing the product from known suppliers, contacting potential suppliers of the products for an estimation of the unit cost to procure from them would yield a more accurate estimate of the total unit cost.

Other Approaches for Costing and Pricing a New Product

Other approaches for costing of new products could use customer input; conduct customer surveys using the approach described in Chapter 9 to find out how much customers may be willing to pay for your product.

The customer survey can provide other valuable input to a start-up business. *The information from the survey will be trustworthy to the extent the survey described the product accurately to the potential target customer.* A prototype or an early version of the actual product would provide the customer with the chance to price it more accurately.

(© Dizain / Shutterstock, 282871847)

Once a survey reveals what the customers are willing to pay for a product, the company can work backwards to estimate the total unit cost for the manufacturer. If it is a consumer product that can be purchased off the shelf or from the Internet, start-up businesses could estimate the total unit cost from the selling price by subtracting estimated costs and profits for intermediaries between the manufacturer and seller as described above.

For example, if we used 100% markup at each stage, if the customer would pay $80 per unit, estimate that the retailer would acquire it for $80/2 = $40 from the start-up company, allowing for a 100% markup by the retailer. Thus, $40 would be the start-up company's per unit target selling price to the retailer. The target total cost for the start-up company would be $20, allowing for 100% markup. If the start-up company could make the product or get it made for $20 or less, the start-up company could be profitable. If not, the company may continue to research their cost structure and pricing. The use of 100% markup by start-up companies for new products with little competition and patent protection is not that rare. Given all the uncertainties associated with a new product of a start-up, 100% markup provides some cushion from failure. This is one approach to estimating total cost and pricing for a new product by a start-up company.

Because these estimates are approximate, the company may make more than one total cost and price estimates using other approaches.

Use Pricing to Compensate for Approximation in Costing

The above section uses generous markups because cost estimation at the business planning stage for a new start-up could be approximate and crude. Therefore, after the best possible total cost estimation of the product, the selling price for the product must be about two times the total cost (that is, 100% markup) to account for errors in costing. For example, if the total estimated cost is $10 per unit, the new start-up must set the selling price for its product at $20 per unit or more.

The challenge then is to investigate if the product could sell for $20 or more per unit. If not, the company will be compelled to lower the cost of the product. For example, if the product is not likely to have a significant market over $15 each, the company may lower the selling price to $15 and must seek to get it made for 15/2 = $7.5 or less. This then is a tangible challenge for the company to change and modify the design of the product, or look for substitution of materials, or consider other suppliers, and so on, until the product can be made or acquired for the lower price. At this stage, business model and business plan are impacted and changed to fit the new total cost and selling price realities.

Example of Cost, Price and Cash Flow estimation

Tables 27.1, 27.2 and 27.3, completed in sequence, show an example of how to estimate annual costs and revenue for a new business. Start with Table 27.1, which requires the business planner to estimate the annual sales in number of units and selling price per unit to complete the table. Table 27.2 shows the estimate of the cost-of-goods-sold. Finally, Table 27.3 shows selling and general administrative expenses.

The data from Tables 27.1, 27.2 and 27.3 are drawn into Table 27.4 to project the cash flow for five years. Read the notes for Table 27.4 to understand how the cash flow is prepared and how to interpret the entries in the table.

The invaluable benefits of preparing these four tables are many. The entrepreneur learns what data to collect, how to collect them, and the key relationships between variables in the tables. In the process, the entrepreneur learns to understand several business issues that could make or break his/her start-up business; this is one of the best learning-by-doing exercise in the difficult journey that an engineer-entrepreneur undertakes. The sincere effort of the engineer-entrepreneur to arrive at Table 27.4 will pay off substantially. It will provide a birds-eye-view of the business and the tight links between key variables that can make or break the business.

(© Keith Bell / Shutterstock,
182001812)

Table 27.1. *Revenue Computation: Example*

Year	1	2	3	4	5
Revenue Estimate					
Units Sold (1)	10,000	12,000	16,000	20,000	24,000
Price Per Unit estimate $ (2)	22.00	23.00	24.00	25.00	25.00
Installation /Service Revenue $ (3)	1.00	1.00	1.00	1.00	1.00
Total Revenue Per Unit $ (4)	23.00	24.00	25.00	26.00	26.00
Total Revenue Per Year $ (4)	230,000	288,000	400,000	520,000	624,000
(Enter in Cash Flow table)					

(1) Enter estimated units produced/sold for five years based on forecasts

(2) Enter estimated price per unit sold

(3) Enter installation and service revenue per unit, if any

(4) Computed (if you use the Excel templates provided on this book's website, it will do the computations for you, and transfer the entries in the last row to the cash flow table, Table 27.4; download template from www.engineer-entrepreneur-book.com)

Table 27.2. *Cost-of-Goods-Sold: Example*

	1	2	3	4	5
Cost Of Goods Sold:					
Units Produced (from Revenue table) (1)	10,000	12,000	16,000	20,000	24,000
Estimate Direct Cost Per Unit:					
Materials 1 $ (2)	7.50	7.50	7.00	6.75	6.75
Materials 2 $ (2)	2.00	2.00	2.00	2.00	2.00
Labor $ (3)	1.00	1.00	1.00	1.25	1.25
Other/Outsourcing $ (4)	0.50	0.50	0.50	0.50	0.50
Cost Per Unit $ (5)	11.00	11.00	10.50	10.50	10.50
Indirect costs	0.00	0.00	0.00	0.00	0.00
(add 0% to 100% of previous row)					
Indirect cost estimated for example: 0%					
Total Cost Per Unit with Indirect Costs	11.00	11.00	11.00	11.00	11.00
Total Cost of Goods Sold $ (5)	110,000	132,000	168,000	210,000	252,000

(Transfer to Cash flow table)

(1) Enter units produced per year from Revenue Table (Table 27.1)

(2) Enter estimated total material cost per unit (two different materials are needed for this product)

(3) Enter estimated direct labor cost per unit

(4) Enter estimated other cost including outsourcing costs for the product

(5) Computed (if you use the Excel templates provided on the author's book website, it will do the computations for you, and transfer the entries in the last row to the cash-flow table, Table 27.4; www.engineer-entrepreneur-book.com)

143

Table 27.3. *Selling and General Administrative (SGA) Expenses: Example*

Units Produced (from Revenue Table 27.1)	10,000	12,000	16,000	20,000	24,000
YEAR	1	2	3	4	5
Selling Expenses Per Unit (1)					
Sales Commissions	0.50	0.50	0.50	0.50	0.50
Sales Salaries	8.00	1.50	1.75	1.75	2.00
Shipping	1.00	1.00	1.00	1.00	1.00
Advertising	0.25	0.25	0.25	0.25	0.25
Other	-	-	-	-	-
Total Selling Expenses Per Unit (2)	9.75	3.25	3.50	3.50	3.75
Total Selling Expenses per year (2)	97,500	39,000	56,000	70,000	90,000
General and Administrative (GA) Expenses (3)					
Salary and Benefits	30,000	50,000	70,000	90,000	100,000
Utilities	1,000	1,000	1,000	1,000	1,000
Rent	2,000	2,000	2,000	4,000	4,000
Insurance	500	500	500	500	500
Other	5,000	5,000	5,000	10,000	10,000
Total GA Expenses (2)	38,500	58,500	78,500	105,500	115,500
Total SGA expenses (4)	136,000	97,500	134,500	175,500	205,500
Transfer to Cash flow table					

(1) Enter estimated selling expenses

(2) Computed

(3) Enter estimated General and Administrative expenses

(4) Computed (if you use the Excel templates provided on the author's book website, it will do the computations for you, and transfer the entries in the last row to the cash-flow table, Table 27.4; www.engineer-entrepreneur-book.com)

Table 27.4. *Cash Flow: Example*

Cash Flow Projections for Five Years Note: Numbers inside parentheses (...) are negative numbers

Major cash flow items	Year 1	Year 2	Year 3	Year 4	Year 5
Total Sales Revenue $**	230,000	288,000	400,000	520,000	624,000
(Cost of Goods Sold) $	(110,000)	(132,000)	(168,000)	(210,000)	(252,000)
Gross Margin $**	120,000	156,000	232,000	310,000	372,000
(Depreciation)*	-	-	-	-	-
(SG&A) $	(136,000)	(97,500)	(134,500)	(175,500)	(205,500)
Operating Income $**	(16,000)	58,500	97,500	134,500	166,500
(Interest Expense)	-	-	-	-	-
Income Before Taxes $	(16,000)	58,500	97,500	134,500	166,500
(Taxes) $*	4,800*	(17,550)	(29,250)	(40,350)	(49,950)
Net Income $**	(11,200)	40,950	68,250	94,150	116,550
Depreciation Add Back*	-	-	-	-	-
Net Cash Inflows / (Outflows) $	(11,200)	40,950	68,250	94,150	116,550
Beginning of the year Cash Balance $	-	28,800	69,750	138,000	232,150
Seed Investment (1)	30,000				
Angel investment (1)	10,000				
(Capital Expenditures)	-				
Loans / (Loan Payments)	-			-	
End of the year Cash $ (2)**	28,800	69,750	138,000	232,150	238,586

* Advanced accounting issues. Beginners may ignore this. Year 1 tax is positive reflecting the fact that when the company begins to make profits, the first year's loss would exempt the subsequent year's profits from taxes (this is advanced accounting practice: do not dwell on this; there is more to learn from this table).

** These are the important rows that an investor and business person will pay attention to. As an engineer-entrepreneur, learn to appreciate why these rows are important. **Total Sales Revenue row** is important because it tells the reader the magnitude of revenue flowing into the company, year after year. **Net Income row** tells the reader, the amount of cash left over after all expenses are paid off and is available to the start-up owner. **End of Year Cash** is the "bottom line." It is one important criterion for investors. As an engineer, understand that by making wise business decisions to alter the rows above the bottom line (i.e., Tables 27.1, 27.2 and 27.3) you can bring about the desired changes in the bottom line.

145

Table 27.4 (*continued*)

(1) Enter Seed and Angel Investment received, expected to receive, or necessary for this project; Year 5 shows that seed investors are paid back $90,000, and angel investors are paid back $20,114 as part of an agreement with the investors; the terms of such agreements vary from one investor to the other – there are no fixed rules; they include equity, loans, royalty, and combinations thereof.

(2) Computed (if you use the Excel templates provided on this book's website, it will do the computations for you; in addition, it will move all relevant numbers from Tables 27.1 to 27.3 – they are Revenue Table, Cost-of-Goods-Sold Table, and Selling and General Administrative Costs Table)

Note 1: In the above table, the End of the Year Cash in Year 5 is $238,586 (from the "bottom line);" this amount is too small to attract serious investors in the United States. Let us assume this small business has targeted only a small piece of the Total Addressable Market (TAM; see Chapter 19). Given that there is still a very large untapped market for the product in the United States, the business plan may be modified to address a larger target market to make the bottom line fatter than it is.

Note 2: In its current financial state, it can sustain a part-time hobby or a small business with one paid employee in the United States. To increase the chance of attracting serious investors, a start-up business must project $1-$3 million or more as End of the Year cash by Year 5, with plenty of room to grow to reach total sales of $10–30 million per year, down the road.

Note 3: Go to the book's website at www.engineer-entrepreneur-book.com to download the Excel template that can produce Tables 27.1 through 27.4 for your business plan. The templates are based on current US accounting practices with built-in formulas for computed values in the tables. Data from Tables 27.1 through 27.3 are automatically posted in the Cash Flow Table 27.4. The template does not come with guarantees or warrantees. It is a helpful tool made available freely to students of this book for quickly developing a five-year cash flow. Although Table 27.4 in not precise, it could be used to **estimate a valuation of the company for investors.**

Note 4: The bottom line, or the "End of the Year Cash" row, is commonly used by investors to estimate the valuation of the company prior to making an investment in the business. Investors would also pay attention to "Total Annual Sales" in the first row as well as the "Income Before Taxes" in making an estimate of the valuation of the company. The valuation of the company determines the percent of ownership in the business an investor would require in return for his/her investment. If a company were valued at $1 million, an investor making a $100,000 investment would ask for 10% equity in the business.

Note 5: Investors use many different criteria for valuing a company prior to investment. One of the simplest tools is to multiply annual sales by a factor to find the valuation for investment. For example, a restaurant business in the middle of the city with annual sales of $1 million a year for the last 5 years may be valued at ($1 million) x M, where M is the multiple for the specific business being evaluated. This no-growth business may deserve M = 1; that is, the restaurant will be valued at no more than $1 million. However, a new technology business with potential sales all over the United States, with annual sales of $1 million but growing at 50% a year for the last four years, may be valued at 5 to 10 times annual sales (M = 5 to 10); so, such a high-growth company may be valued at $5 million to $10 million based on an annual sales of $1 million. Thus, annual growth rate in sales lifts the valuation of a company.

ASSEMBLE A BUSINESS PLAN
TO ATTRACT INVESTORS

28 The Business Model versus the Business Plan

A business model (B-Model) is not a business plan (B-Plan); a carefully selected B-Model becomes the basis for a business plan. If a B-Model were changed, the B-Plan would most likely change substantially. Often start-ups need financing from investors; without an investor some start-ups cannot take off. From an investor's perspective, before the decision to invest, an investor would look for a B-Plan, which is built on one selected B-Model among many possible options.

The ***business plan adds a time dimension to a business model*** – normally a five-year horizon is used in preparing business plans. A business plan enables the marketing of the start-up to potential investors and business partners too. The business plan and business model form a blueprint during the early stages of the start-up. A flawed business model would ensure the failure of the start-up regardless of the attractiveness of the business plan to investors. As the start-up takes shape, the business model and B-plan may need adjustment, and the entrepreneur must be sensitive to the needed changes and act on them when needed.

During the pre-start-up phases described earlier in the book, an entrepreneur must devote time to the selection, testing and refining of the business model before completing the business plan. The financial section of the business plan is a prediction of cash flows and profit/loss over a five-year period assuming a preselected business model is implemented over that period

The Business Model

Phase 3 is a planning phase. During Phase 3, technical and business issues merge. Two important, interdependent outcomes of this phase are the business model and the business plan for a five-year period. A business model is a collection of decisions about the key components of the business, such as:

- Which customers to target? Should we target customers based on age, sex, region, etc.?
- How do we distinguish our product to attract and retain customers?
- How to get the message to the customers; how do we promote the product?

- What kind of sales force do we need?
- Who will make the product – our company or a supplier? Which supplier?
- Should we use distributors, retailers, drop-shippers, or other intermediaries?
- How do we deliver the product to the customers?
- How do we take orders from customers?
- How quickly do we fill customer orders?
- How do we price the product?
- What will be the target profit margins?
- Should we sell or lease our product, or both?
- How do we handle product returns and warranties, if any?
- How to protect our market and customers from competitors?

The Business Plan

The business plan is the anticipated financial result of the business model implemented and functioning over a five-year period. The business plan is a written document to communicate to outsiders and potential future stakeholders such as investors, partners and managers. This document is a living document; it is constantly improved and refined. It can serve as a road map in the initial Execution phase. A complete business plan document may include all of the following:

One-page Executive Summary (see Note 1 below)

1. Introduction
   a. Mission of the organization
   b. Goal of the business plan (communication to investors, partners and managers)
2. The idea
   a. Opportunity, need, or problem
   b. The value to the customer
3. The product
   a. Patent search summary
   b. Status of patents or applications for patents
   c. Description (without compromising patent applications or secrecy needs)
4. The customer
   a. Who is the target customer?
   b. Customer survey
      i. Who was surveyed?
      ii. How many were surveyed?
      iii. Major conclusions based on the survey
5. The Market
   a. Estimate the US market; NAICS data for the market (US Census Bureau, etc.)
   b. Estimate TAM, SAM and target market
   c. Conclusions

6. Competitors
    a. Who are the competitors?
    b. Why customers will buy our products?
    c. What is our product's advantage over competition?
7. Marketing and sales
    a. Channels to be used
    b. Challenges and cost of reaching customers by various channels
    c. How do we get the product to the customer?
    d. How do we sell the product?
8. Costs
    a. Fixed costs
    b. Variable costs
    c. Break-even
9. Pricing
    a. Price estimation
    b. Demand for five years
    c. Break-even analysis
10. Organization
    a. CEO and managers, qualifications
    b. Legal organization
11. Investment
    a. How do we expect to finance the business in the first five years?
    b. How much investment is needed in years 1 through 5?
12. Cash flow and profit statement
    a. Our assumptions
    b. Conclusions based on the bottom line
    c. Why our profit statement should be attractive to a potential investor
13. Statement of Ethics
14. Uncertainties and risks
15. APPENDIXES
    a. Detail description of the product (without compromising patent applications or secrecy needs; see Note 2 below)
        i. Parts, specifications, suppliers
        ii. Technical specification, standards and engineering analysis of key issues
    b. Drawings (subject to secrecy limitations of the company)
    c. Include supplementary information to convince the investor

Note 1: A business plan is a document with a *one-page executive summary*, which includes brief statements about the following to inform and convince potential investors to read the complete business plan:

1. The unmet need in the market underlying the business;
2. Description of the product/service and how it meets the unmet need in the market;

3. The business model (how the company would produce income);
4. Target market, marketing, sales and distribution;
5. Competition in the target market;
6. Production/sourcing;
7. Intellectual property, its ownership, royalty to pay, if any;
8. Management team;
9. A short summary of costs, sales revenue and profits over five years with projected cash flow;
10. The investment needed to operate at the projected sales level and how soon the investment will be recovered by the start-up;
11. Why this is a sound investment.

Note 2: Some companies may require a non-disclosure/secrecy agreement signed by readers before disclosing the full business plan. If your company does not wish to share the entire business plan with potential investors or partners, use an abridged shorter version without compromising sensitive information.

29 The Business Plan: The End of Phase 3

Business Models and the Time Dimension

The financials section of the business plan introduces a time dimension, often five years. Thus, this section estimates all costs and revenues over the time period, while assuming the already chosen business model is true. Since a business plan introduces a five-year projection of sales, costs, and revenues based on the underlying business model, its accuracy is a function of the robustness of the B-Model, the quality of the assumptions and relationships in the B-Model, as well as the accuracy of demand, costs and revenue forecasts for five years into the future.

(© Kheng Guan Toh / Shutterstock, 47603281)

The Internal Rate of Return

The financial metric called the internal rate of return (IRR) may be used by some investors to rank and choose among multiple investment options. It can be defined in more than one manner: technically, it is the discount rate that reduces to zero the present value of all future cash flows (for the five years in the business plan). IRR is also called the economic rate of return (ERR).

Investors with limited capital to invest in start-ups, wanting to choose among several investment options, would tend to pick investments with the highest IRR; the higher the IRR, the quicker the full recovery of the original investment. For example, an investment opportunity with an IRR of 105% promises to return the full investment three times faster than an investment with an IRR of 35% (i.e., 105/35 = 3).

An entrepreneur must assume that investors want to recover their investment as early as possible; therefore, an entrepreneur, while preparing a B-Plan to attract investors, must work towards a B-Plan with a bigger IRR to attract investors.

Note: if the investment is not in a risky start-up business, the return on investment sought by investors is far smaller than for risky start-ups. For example, if the interest rate for a bank-issued Certificate of Deposit (CD) is 2%, safe and guaranteed, a cautious investor would gladly invest in a 6% opportunity that is also safe and guaranteed. Since investments in start-up businesses are not "safe" and returns are not "guaranteed," investors want very high returns per year as a compensation for the risk involved – generally 35% a year or more.

The Criticism of Business Plans

A common and valid criticism of business plans is that the entrepreneur devotes far more time making the B-Plan and devotes inadequate time to the B-Model at the core of that plan. Critics want more effort on developing a robust and validated B-Model before devoting time to developing a B-Plan. The essence of this justifiable criticism **is that** *it is a waste to spend effort and time on a B-Plan with a flawed or underdeveloped B-Model*. Critics believe that the B-Model demands considerable time and effort for developing, refining and testing of the primary assumptions and relationships among variables critical to the success of the B-Plan.

A business plan is the document that signals the end of Phase 3. It moves the technological innovators to Phase 4 and beyond. Its audience is composed of potential investors, who read it or hear its presentation to make up their mind to invest or not to invest in the business. For the innovator, it serves as a flexible document to guide execution in Phases 4 & 5. During the execution of Phases 5 & 6, a business plan is bound to undergo changes to respond to circumstances and to take advantage of new opportunities not anticipated in the original document prepared in the pre-start-up Phase 3.

Sample business plans are offered on the Internet. One such site is BPlans.com.[1] Palo Alto Software sells *Business Plan Pro* software for business planning, priced within reach of young innovators.[2]

---

[1] Bplans.com: www.bplans.com/sample_business_plans.cfm; see also www.bplans.com/samples/sba.cfm
[2] Palo Alto Software: www.businessplanpro.com/template_offer_lt/?gclid=COuEyP3MsrECFUE GRQodfAMASQ&; Rocket Lawyer: www.rocketlawyer.com/secure/interview/questions.aspx? document=12449668&utm_source=103&v=3#q1

30 A Business Plan Is a War Plan: Anything Can Change

The old adage about the best-laid battle plans going to waste once the shooting starts is a tried-and-true statement. This is why a business plan is very much akin to a war plan; it is hard to predict everything that may go wrong before actually beginning operations. But with meticulous, comprehensive planning, including contingency planning, there is more of a chance that unfavorable developments can be countered or overcome quickly; surprises will be fewer.

(© Trekandphoto / Adobe Stock, 64836881)

The need for planning increases as the complexity and uncertainty increase. One of the most-widely publicized examples of meticulous war preparation and its execution is the D-Day invasion of France by the Allied Forces. Planning and preparation for the Allied landing in Normandy on June 6, 1944 took many years. And even with all this planning and preparation, many things were necessary to align just right for the beachhead to be established to ensure the long-term success of the invasion and, consequently, the war as a whole. According to the National D-Day Memorial:

> It is hard to conceive the epic scope of this decisive battle that foreshadowed the end of Hitler's dream of Nazi domination. Overlord was the largest air, land and sea operation undertaken before or since June 6, 1944. The landing included 5,000 ships, 11,000 airplanes, and over 150,000 service men. After years of meticulous planning and seemingly endless training, for the Allied forces, it all came down to this: The boat ramp goes

down, then jump, swim, run, and crawl to the cliffs. Many of the first young men ... faced over 200 yards of beach before reaching the first natural feature offering any protection. Blanketed by small-arms fire and blanketed by artillery, they found themselves in hell [all the years of planning could not foresee or prevent this disaster]. When it was over, the Allied forces had suffered nearly 10,000 casualties; more than 4,000 were dead. Yet somehow due to planning and preparation, and due to the valor, fidelity, and sacrifice of the Allied Forces, Fortress Europe had been breached. (The National D-Day Memorial, www.dday.org/history/d-day-the-invasion/overview)

Business plans are meant to enable the formation of a "beachhead" in the market-place for the start-up. Without a business plan, the odds of success are reduced. *The process of developing the business plan prepares the start-up entrepreneur for possible surprises ahead. The odds of success of the business are improved by a strong business plan; it allows new businesses to weather their own "D-Days" and survive.*

Not every innovator will succeed in developing a business plan that is thorough and complete in all its sections. The business plan, however, is a living document; it evolves and gets better with time and with execution. Whether it is a plan for operating a small start-up business or for a massive wartime operation – all of them change as they encounter reality.

Before the D-Day landing, there was another Allied landing on August 19, 1942, at Dieppe, code-named Operation Jubilee. It was a failure; almost 60 percent of the men were killed, wounded, or captured. But this failure is said to have been instrumental in the superior planning and preparation for the next landing at Normandy in 1944. Business start-ups are no different; learn – from your mistakes if necessary, but better from someone else's – change plans, and execute.

Business plans have launched both mom-and-pop shops and industry giants of today. Here is one story of a good business plan changing the face of the industry as we know it:

In 1994, at age 30, Jeff Bezos came across a report projecting annual Web growth at 2,300 percent. To a Princeton graduate in electrical engineering and computer science, "that was a 'huge wake-up call.' So, three months later, he walked away from being the youngest senior vice president of D.E. Shaw, a Wall Street hedge-fund firm, and set out west in an aging Chevy Blazer with his wife MacKenzie ... By the time they reached Seattle, he'd written the initial draft of a business plan on his laptop, retained a lawyer by cell phone, and started the search for a vice president of development. Five days later they moved into a rented house in the suburb of Bellevue, set up shop in the garage, and voila! Amazon.com, the online retailer that bills itself as Earth's Biggest Bookstore, was born.

(*Success* magazine, July 1998; www.success.com/article/
from-the-archives-jeff-bezos)

Engineer-entrepreneurs will be well advised to use mentors or business partners while preparing business plans to get the needed dose of reality while planning, and to minimize surprises when implementing those plans.

31 Making the Start-Up Business Financially Feasible

In 2010, in the midst of the severest economic recession since the Great Depression, an idea for starting a new angel investors club was proposed by an experienced angel investor in a small US town. In about three months, the new club had forty qualified members. In the first year of its existence, the new club funded six innovations, in participation with other angel investor clubs in a 150-mile radius; the author became one of the founding members.

The experience described in the preceding paragraph is good news for innovators looking for investors in the United States. It is important to remember that *if there is a good and unique product with capable innovators and managers behind it, investment funds will follow. Premature fear of financing a potential new business should not be permitted to hinder innovators in developing their ideas into a viable product during Phases 1 and 2. Serious considerations of financing the business are better delayed until Phases 4 and 5, after the first three phases are completed with rigorous effort.*

One important purpose of this book is to convince potential innovators that it is important to develop their original product idea into a viable, unique, robust, engineered product during Phases 1 and 2. A grave mistake of potential technological innovators who are new to the business world is the belief that no one will fund their idea, which denies the idea the chance to go through the earlier phases described in this book. The first three pre-start-up phases of technological innovation lay the foundation for the funding phases to follow.

The technology and the product combined with a strong management team are very likely to attract private equity in the form of seed money. Given this, the innovator would be better off devoting time to developing the technology and product while delaying the concerns about funding the start-up.

Phase 5 Corresponds to Stages 1–4 in Entrepreneurship Literature

Entrepreneurship and finance (EF) literature recognizes several stages (see the five stages on the left in Table 31.1) in the life of a new entrepreneurial venture from

Table 31.1. *Overlap between Entrepreneurship/Finance Literature and the Seven Phases of Technological Innovation (Chapter 3)*

Entrepreneurship/Finance (EF) literature	Seven phases of technological innovation
	Phase 1: idea refinement*
	Phase 2: product development*
	Phase 3: business plan*
Stage 1: start-up funding	Phase 4: start-up funding
Stage 1: start-up execution	**Phase 5: start-up**
Stage 2: survival	**Phase 5: survival**
Stage 3: rapid growth	**Phase 5: rapid growth**
Stage 4: IPO or buyout	**Phase 5: IPO or buyout**
Stage 5: Maturity	Phase 6: Maturity
	Phase 7: Continued growth, innovation induced

* This book is focused only on the first three phases listed in the first three rows; these phases are not covered in the Entrepreneurship/Finance literature.

** The five stages in the left column overlap with Phases 4–6 of the Seven Phases of Technological Innovation.

the time of start-up. In the seven-phase view presented in Chapter 3 of this book, Stages 1–5 of the EF literature coincide with Phases 4–6 of the seven phases presented in this book. Phases 1–3 of technological innovation (Rows 1–3 in Table 31.1) are generally not included in the EF literature. This book serves the needs of engineers and scientists, who need Phases 1–3, where their contributions are critical to the success of the start-up business; however, they are not likely to find this information in the EF literature.

According to the EF literature, financing of start-ups occurs in a structured manner as shown in the following list (the range of funding at each stage is approximate; it can vary from one firm to the other):

1. Seed funding – early idea-proving and product development stage; sources of funds are own, family funds, or private equity for $0–200,000; sometimes called Round A financing.
2. Start-up financing – initial production and sales begin; sources of funds may be own, friends, angel investors and private investors for $200,000–3 million; Round B-C financing that follow Round A.
3. Survival and growth stage – demand and sales grow; sources of funds are small group of private investors, venture capitalists, internally generated cash by business operation, and investment bankers in the $3–10 million range.
4. Maturity stage – business is stable, established; sources of funds are initial public offering of stocks (IPO), bank loans, bonds, commercial banks for funds beyond $10 million. An IPO event occurs when the company makes ownership available to the public by listing the company on a stock exchange; it enables the company to sell part of the ownership to the public to raise capital.

Angel Investors and Venture Capitalists

In the United States, it should be noted, if you have a good product, a good business model, and good people to run the start-up, private capital (or equity) will chase you. The author has observed that there is plenty of private equity money eager to fund worthy technological innovations in the United States; the *Shark Tank* TV show is the most visible example of this. Thousands who became wealthy during the dot-com era are eager to invest their fortunes in the next technological innovation by funding worthy projects as angel investors or as venture capitalists.

Angel investors, who are wealthy private investors and qualified according to SEC guidelines (must have high income and assets), invest their own money, whereas venture capital firms invest for others from pooled funds of investors under the control of the venture capital firm. **Venture capital firms** often invest in a business years after angel investors, if the business does well and has a promising future; that is, *they are willing to take lesser risks than angel inventors.* Consequently, angel investors expect high annual returns from their investment – 30% to 200% or more per year of investment.[1]

Venture capitalists, who follow angel investors if the start-up is successful, would require an ownership share and board membership in return for their investment. Typically, they prefer to sell their ownership share in three to seven years later; their expectation for return is high and they target high growth companies that have emerged from the start-up stage with the wind at their backs.[2]

The Role of Banks

Investment bankers play a key role in an initial public offering (IPO), when a successful start-up converts from entirely private ownership to public ownership by listing on a stock exchange. Commercial banks play a role in providing working capital (loans) for an ongoing business that has customers and stable revenues but needs temporary funding to run the operations. Working capital is similar to credit cards used by individuals; the company makes periodic payments to the bank to cover the loan.

US Small Business Administration

US Small Business Administration (SBA) plays a role in getting bank loans for start-ups by providing government loan guarantees to banks that lend to start-ups. Visit www.sba.gov for more information.

A final note: This is a bird's eye view of financing a start-up. IF you reach this stage, speak to multiple investors and seek legal advice.

[1] www.nbai.net
[2] www.vfinance.com; www.nvca.org

32 What Angel Investors Look for in a Phase 4 Company Seeking Funding

To supplement any personal funds or funds from family and friends, during Phase 4, start-up capital may be solicited from wealthy private investors. The start-up that is seeking funds must get paperwork prepared by a law firm because, in return for the investment, owners give up a part of the company in a legally binding transaction subject to rules and regulations of the Securities and Exchange Commission (SEC) for businesses operating in the United States.

Start-ups may obtain funds in sequential steps from *investors*, often called Series-A, -B, -C, and so on. After the funds from an earlier funding Series is exhausted, start-ups go for refunding until sales revenue makes the company financially self-sustaining. SEC-approved private investors participate in these series of private funding. They must be wealthy, with SEC-specified minimum asset base and/ or annual income to withstand total loss of the investment in a start-up, and must acknowledge this risk before participating in the investment.

(© Master_art / Shutterstock, 182881166)

In the following example, a start-up was issuing (read: selling) Series-A Preferred Stock in the first round of private funding by offering shares to a few investors outside the company in a legally binding manner in return for partial ownership of the start-up. In this case, Series-A preferred stock was convertible into common stock if the company is sold or goes to an IPO.

Series-A round is important to new companies. A typical Series-A round could be in the $200,000 to $10 million range in exchange for 20% to 49% of the company. Series-A could finance the company for up to the first two years as it develops products, begins marketing and branding, hires its initial employees, pays some professional fees to law firms and performs other early start-up activities.

Because start-ups are not listed on public stock exchanges, owners approach known angel investment networks to reach potential qualified private investors and angel investment clubs. Even during the economic recession of 2008–2012, these clubs were active in funding start-ups while banks restricted lending to small businesses in general. To raise capital from private investors, a company's founders and leaders will make presentations to interested individuals or one or more angel investor clubs.

A Real Case of a Technological Start-Up Seeking Funds

XYZ Company, based on the East Coast of the United States, was a start-up company that was looking for funding in early 2012.[1] An angel investment club somewhere in the United States read the business plan submitted by the company and invited the principals to make presentations to their members, who might invest through their club.

One of the angel investor club's leaders prepared the following due-diligence report for their members.

Evaluations of XYZ Company for Club Members (Partially Reproduced)

BACKGROUND: Incorporated in 2010, this start-up had a board of directors, an experienced CEO and top officers, and engaged a law firm that prepared the legal documents for Series-A funding. The company was in Phases 4–5. The company was evaluated on the following fourteen items by the angel club leaders to determine its suitability for investment.

1. **Overall assessment of the company:** Location of the company, its intended products for the medical devices industry, patents applied for, licensing agreement to use certain technologies invented at a university, the funding source for the research completed at the university, the date of the original research grants to the university, start date, how long the start-up had been in existence, and current status. The product was a virtual interactive system.
2. Revenue projection for five years including pretax profit for each year and capital needed by the company.
3. **Management of the company:** CEO was a successful health care technology entrepreneur with a successful start-up named ABC in a large city in Georgia. Before ABC, he was with a national consulting firm. Founder and Chief Scientific

---

[1] Real names of involved parties are withheld.

Officer (and inventor) was a surgeon trained in a famous medical clinic. He developed a computer image technology that was licensed to a well-known company. He cofounded DEF Company, which went public in 2005.

4. Products were described and evaluated.
5. **Markets and marketing:** the device had applications in the medical device industry but could have applications outside this industry in industrial training. It could have potential military application in the field of remote communication.
6. **Competitors:** No competitor offered patent-pending features of this company's remote communication technology.
7. Competitive advantage evaluated.
8. Intellectual property evaluated.
9. Valuation of the company and terms of the offer of ownership in return for the investment were evaluated.
10. **Product potential:** The product had immediate application in surgical hospitals and the military.
11. **Financials:** The equipment's selling price to hospitals and the military, and software maintenance fee per month per piece of equipment sold were evaluated.
12. **Exit strategy for investors:** Exit likely to occur through acquisition by one of the larger companies in the medical equipment/software industry or a communication company. Exit options evaluated.
13. **Risks:** This industry is slow to adopt new technology, and the equipment, being expensive, must go through a slow approval process in a hospital.
14. **Evaluation of the investment in the company so far:** founders have already invested $200,000 in the company and will invest another $50,000 in this current Series-A solicitation. The university licensing the technology to the start-up has invested $120,000, and the founders have received no income from the company for the two years the company had been around.

The above list of items evaluated by the leaders of the angel investment club provides a valuable lesson on what a well-organized angel investor looks for, and how technological innovators may prepare for funding in Phases 4 and 5. This angel investor group invested $300,000 within 30 days of the presentation. Notice the list of items carefully evaluated based on information provided in the business plan, legal documents, and the presentation. Compare this with the list of items suggested for inclusion in your business plans in Chapter 28. Investors pay attention to what you say in your business plan.

Note that the above list contains only non-proprietary information that could be disclosed; all proprietary information including the details of the company, the angel investment club, the product, and so on are left out. Still, it conveys to the reader what seasoned investors look for while investing in new start-up businesses.

Question: What did you learn from this case?

33 Ethics in Engineering and Business Professions

What Is Ethics?

Ethics refers to standards adopted voluntarily by groups or professions that everyone in the group *wants or expects* others in the group to follow. Ethical behavior includes obedience and conformance to all laws of the nation, state, or city. Organized groups often have internal codes of conduct spelling out the ethical standards for all group members; members of the group may lose their group membership or privileges for violating the group's code of ethics.

Laws of the nation, state, or city tell us what is right and wrong from the perspective of the lawmakers. Ethical behavior calls for obedience to or observance of the law both in its spirit and letter. Ethics of a group may go beyond the laws of the nation when it observes a more stringent moral code that is not covered by the laws of the nation. For example, although alcohol consumption is legal in the United States, one may choose to give it up as part of a moral code of his/her religion, or of a private association. Unethical behavior is either:

1. Law-breaking behavior; or
2. Behavior that contradicts professional, religious, spiritual or moral standards, or an employer's code of conduct.

Consequences of Unethical Business Decisions

The conduct of bankers and the mortgage industry employees before 2008 caused significant losses to both private and institutional stock market investors. Thousands of homeowners lost their home values or even lost their homes altogether to foreclosure, in addition to losing their creditworthiness. Furthermore, the conduct of banking and insurance company employees led to a deep recession, and millions in the United States lost their jobs and/or their wealth, causing widespread hardships. Some of the actions by bankers and mortgage industry employees might have been illegal. But it is an absolute certainty that some of their actions were unethical business practice, such as, for example, approving mortgages for people who clearly had no means to repay them.

Unethical behavior is a choice. One does not have to be unethical to succeed. A technological innovator faces ethical choices all the time. It is recommended that a start-up business choose an ethical track from inception to save the business and its employees from inevitable disasters in the future, and for the enjoyment of those who work for the firm and with the firm.

From a purely practical perspective, an inventor cannot invent much, if anything at all, if incarcerated for unlawful behavior. It is equally damaging to the start-up's business potential if its founders, officers, or employees are publicly found to have been involved in unethical conduct. Lawful and ethical behavior is thus good for your business and your personal freedom. There are other reasons to act ethically, too; let us consider them next.

Religious, Spiritual and Moral Standards

Many among us are guided by moral, spiritual or religious commitments that hold us to high standards of conduct and require us to live ethical and law-abiding lives. The Bible represents one of the oldest and most followed ethical codex in the world. Consider just a few of its tenets:

1. "Love your neighbor as yourself" (Matthew 19:19; Matthew 22:39; Mark 12:31; Mark 12:33; Luke 10:27; Romans 13:9; James 2:8) – that is, do no harm to others.
2. "Submit to authority" (Hebrews 13:17; 1 Peter 2:13) – that is, obey the laws of the land.

Ethical Decision Making

A practical way of making ethical decisions is to test the decision options before making a decision and discard the options that seem unethical. Several would agree with the following list of tests (adapted from Michael Davis, 2003).[1] Discard the option if it fails one of the tests:

A. **Harm test:** Does this option do harm to others?
B. **Publicity test:** Would I want my choice made public?
C. **Defensibility test:** Could I defend this choice before my peers, parents or a congressional committee?
D. **Reversibility test:** Would I like to experience the consequences of this choice?
E. **Virtue test:** What would I become if I made this choice, or what would others conclude if I made this choice?
F. **Professional test:** What would my professional ethics committee say if I chose this option?

---

[1] Davis, M. 2003. What can we learn by looking for the first code of professional ethics? *Theoretical Medicine and Bioethics*, 24(5): 433–454.

G. **Colleague test:** What would my colleagues say?
H. **Organization test:** What would my organization's ethics officer or legal counsel say about this choice?
I. **Moral leaders test:** What your parents or your religious leader such as a pastor say?
J. **Role model test:** Could this make me a negative role model for the young and those who look up to me?

If your choice fails one or more of the above tests, it is likely unethical or borderline unethical; don't do it! Say goodbye to the offending choice. Three codes of ethics are presented in the supplements that follow to give the reader a good idea of what is expected by professions and corporations. A start-up business will be well advised to include a code of ethics in its business plan. The following samples could serve as a good starting point.

SUPPLEMENT 1

ABET Code of Ethics for Engineers

(Accreditation Board for Engineering and Technology (ABET) is a non-governmental organization that accredits more 700 engineering colleges or universities in the United States and nearly 30 other countries as of 2016 if they meet the organization's standards)

THE FUNDAMENTAL PRINCIPLES (reproduced)

Engineers uphold and advance the integrity, honor and dignity of the engineering profession by:

I. using their knowledge and skill for the enhancement of human welfare;
II. being honest and impartial, and serving with fidelity the public, their employers and clients;
III. striving to increase the competence and prestige of the engineering profession;
IV. supporting the professional and technical societies of their disciplines.

THE FUNDAMENTAL CANONS

1. Engineers shall hold paramount the safety, health and welfare of the public in the performance of their professional duties.
2. Engineers shall perform services only in the areas of their competence.
3. Engineers shall issue public statements only in an objective and truthful manner.
4. Engineers shall act in professional matters for each employer or client as faithful agents or trustees, and shall avoid conflicts of interest.
5. Engineers shall build their professional reputation on the merit of their services and shall not compete unfairly with others.
6. Engineers shall act in such a manner as to uphold and enhance the honor, integrity and dignity of the profession.

7. Engineers shall continue their professional development throughout their careers and shall provide opportunities for the professional development of those engineers under their supervision.

SUPPLEMENT 2

Corporate Code of Ethics: AT&T Inc.

(An abbreviated version of the public statement on the company website, July 15, 2012; www.att.com/gen/investor-relations?pid=5595)

The Board of Directors of AT&T Inc. (with its subsidiaries, the "Company") has adopted this Code of Ethics (this "Code") to:

- encourage honest and ethical conduct, including fair dealing and the ethical handling of conflicts of interest;
- encourage full, fair, accurate, timely and understandable disclosure;
- encourage compliance with applicable laws and governmental rules and regulations;
- ensure the protection of the Company's legitimate business interests, including corporate opportunities, assets and confidential information; and
- deter wrongdoing.

I. **Honest and Ethical Conduct**

   Each director, officer and employee owes a duty to the Company to act with integrity. Integrity requires, among other things, being honest and ethical. Each director, officer and employee must:

   - Act with integrity, including being honest and ethical while still maintaining the confidentiality of information where required or consistent with the Company's policies.
   - Observe both the form and spirit of laws and governmental rules and regulations and accounting standards.
   - Adhere to a high standard of business ethics.
   - Accept no improper or undisclosed material personal benefits from third parties as a result of any transaction or transactions of the Company.

II. **Conflicts of Interest (details not reproduced here)**

III. **Disclosure**

   Each director, officer or employee, to the extent involved in the Company's disclosure process, including without limitation the Senior Financial Officers, must:

   - Not knowingly misrepresent, or cause others to misrepresent, facts about the Company to others, whether within or outside the Company, including to the Company's independent auditors, governmental regulators and self-regulatory organizations.

IV. **Compliance**

   It is the Company's policy to comply with all applicable laws, rules and regulations. It is the personal responsibility of each employee, officer and director to adhere to the standards and restrictions imposed by those laws, rules and regulations in the performance of their duties for the Company, including those relating to accounting and auditing matters and insider trading.

V. **Reporting and Accountability**

VI. **Corporate Opportunities**

VII. **Confidentiality**

VIII. **Fair Dealing**

We have a history of succeeding through honest business competition. We do not seek competitive advantages through illegal or unethical business practices. Each employee, officer and director should endeavor to deal fairly with the Company's customers, service providers, suppliers, competitors and employees. No employee, officer or director should take unfair advantage of anyone through manipulation, concealment, abuse of privileged information, misrepresentation of material facts, or any unfair dealing practice.

IX. **Protection and proper use of company assets.**

SUPPLEMENT 3

Corporate Code of Conduct: Google

(Abbreviated by the author from Google's public statement on the company website, July 15, 2012; http://investor.google.com/corporate/code-of-conduct.html)

Preface

"Don't be evil." Googlers generally apply those words to how we serve our users ... But it's also about doing the right thing more generally – following the law, acting honorably and treating each other with respect ... The Google Code of Conduct is one of the ways we put "Don't be evil" into practice.

So please do read the Code, and follow both its spirit and letter, always bearing in mind that each of us has a personal responsibility to incorporate, and to encourage other Googlers to incorporate, the principles of the Code into our work.

No Retaliation

Google prohibits retaliation against any worker here at Google who reports or participates in an investigation of a possible violation of our Code. If you believe you are being retaliated against, please contact Ethics & Compliance. The Code of Conduct is elaborated under the headings and subheadings below:

I. Serve Our Users with:
1. Integrity
2. Usefulness
3. Privacy, Security and Freedom of Expression
4. Responsiveness
5. Action when necessary

II. Respect Each Other through:
  1. Equal Opportunity Employment
  2. Positive Environment
  3. No Drug and Alcohol abuse
  4. Safe Workplace
  5. Dog Policy

III. Avoid Conflicts of Interest in the Following:
  1. Personal Investments
  2. Outside Employment, Advisory Roles, Board Seats and Starting Your Own Business
  3. Business Opportunities Found Through Work
  4. Inventions
  5. Friends and Relatives; Co-Worker Relationships
  6. Accepting Gifts, Entertainment and Other Business Courtesies
  7. Use of Google Products and Services
  8. Reporting

IV. Preserve Confidentiality of:
  1. Confidential Information
  2. Google Partners
  3. Competitors/Former Employers
  4. Outside Communications and Research

V. Protect Google's Assets in:
  1. Intellectual Property
  2. Company Equipment
  3. The Network
  4. Physical Security
  5. Use of Google's Equipment and Facilities
  6. Employee Data

VI. Ensure Financial Integrity and Responsibility in:
  1. Spending Google's Money
  2. Signing a Contract
  3. Recording Transactions
  4. Reporting Financial or Accounting Irregularities
  5. Hiring Suppliers
  6. Retaining Records

VII. Obey the Laws on:
  1. Trade Controls
  2. Competition Laws
  3. Insider Trading Laws
  4. Anti-Bribery Laws

VIII. Conclusion

In the above list, items I to VIII are ONLY the headings of sections in the Google Corporate Code of Conduct in 2012. The paragraphs describing each item are not included here. The list shows to the reader that corporate ethics covers a lot of ground.

34 Business as a Legal Entity in the United States

Attorney Services

A start-up business does not need an attorney to form a corporation nowadays when Internet-based self-help for incorporation is accessible through legal documentation services such as LegalZoom and others in the USA. However, when part of the company is being sold to private investors during or after incorporation, doing so will require paperwork that may need legal input from attorneys to ensure that the paperwork is sound and meets Securities and Exchange Commission's (SEC) requirements. In any case, individual owner(s) must decide whether or not they need an attorney's services at incorporation based on the complexity of obligations of the owner and the business to investors, and for tax advantages.

Incorporation (Federal)

In the United States, an incorporated business has legal protection and certain rights. An incorporated business shields the owners from liabilities incurred by the business and may help them reduce taxes. The US Internal Revenue Service (IRS) allows corporations to be either a C Corporation or an S Corporation.

C Corporation. This is an independent legal entity owned by shareholders – the corporation does not pass its liabilities through to shareholders. It is complex to set up and has much paperwork with high administrative costs. Therefore, it is fit for larger, established businesses. It can sell ownership by offering shares in the company via a stock exchange. The C Corporation pays taxes on profits directly to the IRS, and the dividends to owners are subject to taxation as part of the individual's tax returns; this double taxation is a negative. The corporation is taxed using a corporate tax rate, which could be lower than individual tax rate. It is easier to raise capital from investors for C Corporations and it could be more attractive to potential employees. It could have unlimited number of investors.

S Corporation. Primarily, an S Corporation avoids double taxation faced by a C Corporation because all profits and losses pass through to the owners, and the business is not taxed. It is a legal entity that limits the legal liability of owners. The IRS must qualify the S Corporation at formation. By way of comparison, while the legal liability company (LLC, introduced in the next section) is also limited in liability, for tax purposes LLC is treated as an S Corporation. Advantages of an S Corporation include tax savings and its existence as an independent legal entity. Its disadvantages are complex legal requirements and closer IRS scrutiny.

Over the years, S Corporation has been preferred by many small business owners. S Corporation may not have more than 100 shareholders, who all must be US citizens or legal residents. The corporation must hold annual meetings and keep corporate minutes.

Limited Liability Company (State)

A limited liability company (LLC) is a business formed under state law that allows some of the benefits governing a corporation and a partnership. LLC is relatively recent in origin in the US, Wyoming being the first state to enact an LLC in 1977; other states have followed the lead by Wyoming. The owners of an LLC are generally shielded from liabilities arising from company debts, similar to a corporation. LLC is not liable for taxes because profits and losses are "passed through" to LLC owners for tax purposes. LLCs are not required by law to hold formal meetings.

LLC is becoming a more common form of business incorporation in the United States. It offers a mixture of limited liability to the individual/owner and tax efficiencies with the flexibility of a partnership. Owners are members, who can be one or more individuals, corporations, or other LLCs. LLCs are not taxed, but members are taxed on the profits or losses on the individual tax returns just as owners of partnerships.

LLC needs a business name that is unique and approved by the Secretary of State or similar agency in the state where it operates. It needs an operating agreement. Its benefits include limited personal liability, less record keeping than for an S Corporation, less paperwork, and flexibility on sharing profits among members based on member agreement.

Among the disadvantages is the fact that the business may be dissolved when one member leaves unless the operating agreement has provisions to prolong the life of the business. Members also incur self-employment taxes in addition to other taxes.

DBA or Sole Proprietorship

For individuals running a business themselves, a sole proprietorship under a company name requires a DBA ("doing business as") filing with the state or county in the US. This form of business is simple, common, unincorporated business of an individual with no legal separation between the individual and the business; all profits are the individual's to report on the income tax return, and the individual is also responsible for all debts, losses, and liabilities. Benefits include easy tax preparation and paperwork. The main disadvantage is the unlimited nature of liabilities; all business liabilities become personal liabilities if the business legally fails to satisfy the liabilities. This is not a recommended form of business if you need to raise investment capital from others.

Partnerships

This is appropriate when a few partners come together as a business. There are three kinds of partnerships to consider.

General Partnership is one where profits, liability, and management responsibilities are equally divided. The partnership agreement documents may spell out unequal distribution, if needed.

In the US, the partnership must be registered with the state through the Office of the Secretary of State. It will need an approved unique business name, licenses, and permits. Taxes are paid by individual partners for their share of profits; no taxes are paid by the partnership. Partners are not employees (no W2 IRS forms) but will be furnished Form 1065 Schedule K-1 for tax obligations. This is easy to set up and inexpensive to operate. Partners share financial commitment. Partnership may be offered to employees as an incentive for performance and contribution to the business after proven results by the employee. It results in joint and individual liabilities resulting from the partnership. All decisions are coordinated with all partners, who share in all profits.

LP or Limited Partnership is a type of partnership made of two categories of partners: general partners who manage the business, and limited partners who are investors with no management responsibilities. Limited partners have limited liability. Limited Partnership is fit for investors in short-term projects.

LLP or Limited Liability Partnership. In an LLP, partners have a voice in managing the firm without sharing liabilities. Generally, a partner in an LLP is not responsible for the liabilities of the other partners in the LLP.

In summary, the legal entity that a start-up chooses to become will affect the company's and owners' tax obligations to the federal, state and local governments. Additionally, the legal entity affects the number of owners permitted, and the liability of the owners and/or its partners. In recent years, the LLC is the preferred option for a start-up business.

A final note: This is only a bird's eye view of the issues. If you have reached this stage and do not understand the issues, seek legal help.

35 Pre-Start-Up Business Organization and Management

Businesses organize internally by assigning authority and responsibilities to individuals, departments, and divisions. A start-up company's business plan may include the pre-start-up organization and its leaders, including the technological innovators. It is important for those who invest in the start-up to be convinced of the ability and experience of the top management team described in the business plan. In Phase 4, the quality of the innovators and top management will have a positive impact on the investors, who bet their money on a risky start-up.

Organizations are structured such that authority often flows from the top. At the very top is the board of directors (BOD), which makes long-term policy decisions for the organization. These policies are then carried out by the top management or executives of the company composed of the chief executive officer (CEO) or president, and other "C" officers of the corporation such as chief financial officer (CFO) or vice president (VP) of finance, chief technical officer (CTO), chief of operations or chief operating officer (COO) or vice president of operations, chief of marketing or vice president of marketing, and so on. Figure 35.1 is a generic organization chart showing two levels of sample titles: president and VP levels. The figure shows room for two more unnamed VPs. VPs would have lower-level managers reporting to them (not shown in this chart).

Figure 35.2 shows an organization chart for a small start-up company; it often has sales, marketing, and operations functions in the early years; the Founder and

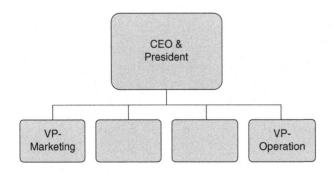

Figure 35.1. Organization Chart for XYZ, LLC

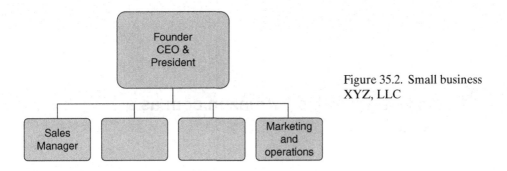

Figure 35.2. Small business XYZ, LLC

CEO may carry many diverse responsibilities in a small start-up. Figure 35.2 has room for two more managers, as the company grows.

Some companies have many layers of employees from the top to the bottom, while others have a "flatter" organization with fewer vertical levels. Firms known for technology innovation tend to have flatter organizations, with the upper management more open to ideas, initiatives and changes emerging from below.

An organization hires managers, staff, and other employees to carry out the annual plan of the corporation to meet planned demand for the products and services of the company. The functions of a manager include planning, communicating, organizing, staffing, leading and controlling. Their skills would include technical as well as the ability to lead, motivate, and influence others.

Managers are responsible for the work of the departments or divisions under their management, and are responsible for the goals, targets and quotas. Successful managers motivate their employees to peak performance with creativity and enthusiasm. Managers are responsible for making a financial budget for the areas under their control and must learn to live under the budget. Overall, managers must meet or exceed organizational goals and objectives.

Staffing the start-up should be planned carefully. During the dot.com bubble of the late 1990s, when private equity funds were coming forth very easily, many start-ups were overstaffed with highly paid IT professionals without adequate work for them. The rate at which a start-up uses up its shareholders' capital is called "burn rate." Once shareholders' capital is exhausted, the company must seek additional funding, or make profits, or shut down. Quite a few new IT start-ups during the late 1990s were shut down because they did not control the burn rate, or the burn rate was way more than necessary. When the burn rate in start-ups is excessive, it would consume the invested capital sooner than necessary and may slow down product development, marketing, and product launch, all vital to the success of the new company. In the more recent case of one start-up business in Atlanta, it received $500,000 Series-A funding from angel investor clubs in the region in the summer of 2011 but could not launch its product as planned in the spring of 2012. The company was seeking additional funding from the same angel investors in the spring of 2012. This time around the angel investors were critical of the company's poor project management, poor personnel policies and a higher-than-necessary burn rate. Consequently, only a few prior investors participated in this round of financing but with tighter control on the company and with a demand that the CEO be replaced.

36 An Illustrative Case: Amazon.com as a Start-Up

In June 2016, Amazon.com was a successful and established worldwide leader in its industry with annual revenue in excess of $107 billion (2015), a market value in excess of $340 billion, and more than 200,000 employees. The company's early days make a useful learning tool for new engineers/scientists. It is easy to get information on the early days of this company and its founder Jeff Bezos because they were the subject of numerous articles and news stories. An article in the *Success* magazine, published in July 1998, provides the following information about the founder and his company.[1]

Jeff Bezos, 1994, Age 30

1. Princeton graduate in electrical engineering and computer science.
2. Motivated by and reacted to the report that annual Web growth was projected at 2,300 percent.
3. This growth rate gave him a sense of urgency.
4. Three months later, on June 30, 1994, he quit his job as senior vice president at D.E. Shaw, a hedge fund, and moved to Seattle in five days.
5. While driving from New York to Seattle with his wife, he put together on his laptop a draft business plan for a company that would eventually become Amazon.com.
6. In less than a week upon arrival in Seattle, he retained a lawyer and started a search for a VP; Amazon.com was born in the garage of his rented home near Seattle.
7. Four years later in 1998, the office still had a unfinished, temporary look – office tables were improvised by 2x4 lumber supporting old doors – the company had sales revenue but was still not generating profits.

Business Plan Outline for Amazon.com

1. List of twenty products that could be sold on the Internet – the business idea.
2. Ranged from clothing to gardening tools.

---

[1] www.success.com/article/from-the-archives-jeff-bezos

3. Thorough research on top five items to sell on the Internet: CDs, videos, computer hardware, computer software, and books.
4. Evaluated each product using a set of criteria including:
   a. Size of the relative markets; books had an $82 billion worldwide market.
   b. Products with low price; since Internet shopping was new, the size of the order should not be financially threatening to new users of this channel.
   c. Range of choice; 3 million books and 300,000 CDs; the computer could be used to organize products with a bigger range of choices, such as books.
5. Location of selling operation chosen from Portland, OR, Lake Tahoe, NV, Boulder, CO, and Seattle, WA.
6. Location to be in a state with favorable sales tax, large high-tech workforce, and near a book-distribution center.
7. Started the business in July 1995 out of a garage in a private residence in the Seattle area.

Key Activities

1. Jeff Bezos, his wife, and three others worked full time with shippers, wholesalers, investors, and software developers.
2. In 1995, the first year of operations, Bezos would take the books to the post office in his car for mailing.
3. In 1998, large post-office-owned semis went to the Amazon warehouse to be filled each day because of the incredible 74 percent growth rate per quarter.

Lessons According to Jeff Bezos

1. Ideas are easy but execution is hard.
   1a. We can scribble 100 ideas on the chalkboard, but the hard part is making them work.
2. Keys to success:
   2a. Thoroughly evaluate the market.
   2b. Bring together a talented and diverse group of people.
   2c. Attract good people; offer them the opportunity to build something important, improve customers' lives, and change the world.
   2d. A great number of employees became owners through generous stock-option incentives.
   2e. Consistently articulate the vision of what is to be achieved; the vision in 1998 was: "Our job is to accelerate access to things that inspire, educate and entertain."
   2f. Figure out what your customers want and then deliver it better than others – this is the only way to succeed, even though it is very hard to execute.
   2g. Amazon customers want selection, ease of use, convenience, and good pricing – "so we are making sure we offer them the best of all those."
   2h. Prioritization; stay focused on the Keys to Success.

EXECUTION PHASES 4–7

THE VIEW BEYOND THE BUSINESS PLAN
CREATING AND OPERATING THE BUSINESS

37 Execution Phases 4 to 6

This book's focus is on Planning Phases 1–3, which are described in detail in the preceding chapters. The four Execution Phases, from 4 to 7, are not fully detailed in this book; however, selected short case studies of firms in Phases 4, 5 and 7 are included in later chapters. This chapter provides a short summary of Phases 4–7 to give the reader an overview of what lies beyond Planning Phase 3.

(© Alexmillos / Adobe Stock, 54371031)

Phase 4: Find Investment Capital

With the completion of the pre-start-up phases, finding investment capital (estimated time 1–24 months) may be a distinct and significant phase for some start-ups. Furthermore, in some start-ups, Phase 5 (creating a functioning start-up) may occur simultaneously alongside Phase 4.

During this phase, the innovator may put together a management team gradually, with minimal salary commitments; form a board of directors with room for investors; form the new start-up with a legal status such as LLC; use legal help to prepare paperwork if investment is solicited from others in return for partial ownership of the company; communicate the business model and business plan to potential investors through meetings and presentations to family, friends, early stage investors/angel investors, and others; revise the business plan as input comes in from investors and potential investors, and refine the business plan as needed with inputs from investors and new top management team members coming in.

Phase 5: Create a Functioning Start-up

Assuming at least part of the investment has come in, the business will get deep into execution in this phase (estimated time 3–30 months). During this phase, a number of strategic and tactical decisions must be made, and expenses (often significant ones) incurred, while any substantial revenue may still be in the future. Ideally, revenue comes in before all the investment is used up. The decisions, implementation of decisions, and expenses during this phase may include the following, although not necessarily in this order:

1. Facilities needed: build or rent (incur expense).
2. Establish staffing, functions, responsibilities, subcontractors (incur expense):
   a. Order processing and fulfillment system;
   b. Marketing, sales;
   c. Production/procurement;
   d. Personnel recruitment and policies;
   e. Accounting/finance; tax and legal matters;
   f. Government permits; meet local, state/federal regulations.
3. Produce goods and make sales, and produce revenue.
4. Set up research and development, and continue product development.
5. If sales are strong, scale up the business promptly, bring in professional managers and move to Phase 6 (Yahoo! and Earthlink experienced surprisingly strong sales instantly, they are exceptions).
6. If the business is scaling up as expected or better, develop a new five-year cash flow projection to manage the growth, and move to Phase 6.

Phase 6: Grow to Financial Steady State or Maturity

Most businesses we come in contact with in our daily life are in either Phase 6 or Phase 7. Financial steady state/maturity may arrive 50–100 months after the inception of a start-up business. Before arriving at this phase, the business must deal with growth.

A new business reaches maturity when growth in revenue slows down followed by end of growth, and eventually experiencing declining revenues. A mature business must make strategic decisions in a timely manner to prevent the decline in total revenue to the business by improving its products, and/or by seeking out newer markets not explored before, and/or by blunting competitors' product appeal, and/or by launching new products. When a start-up, after many years, learns to keep revenues steady or manages to grow, the business has reached a healthy steady state of operation.

The steady-state or maturity phase arrives after a prolonged time of continued growth, when growth must be managed. Failure to manage growth has caused businesses to fail. Growth management must occur in the following areas of a growing business:

1. Output;
2. Support facilities;
3. Sourcing of parts, components and sub-assemblies;
4. Supply chain;
5. Staffing for research, production, marketing, sales, order-taking, etc.;
6. Customers relations;
7. Capital to enable growth;
8. R&D effort and product development;
9. Efficiency of the operation as the business grows.

During the time of growth or upon reaching maturity, the innovators, who started the business, may sell the company for a good financial return to all original investors; for the founders/innovators who started the company, this signifies the end of their entrepreneurial journey (at least until the next idea and the next start-up).

Alternatively, the owners may launch an IPO to make the business a public company, and list it on the stock exchange with help from an investment banking company; among the best-known examples of the companies going public while their founders/inventors continued in leading capacities are Yahoo!, Facebook, and Google.

If never sold, and no IPO is issued, the business will remain a private company under the control and ownership of the original innovators and investors.

38 Phase 7: Six Case Studies of Mature Firms

Many businesses, upon reaching maturity, may become weak innovators, thereby stunting their growth and becoming vulnerable to competition, if not failing outright. Decades after its founding, Apple Inc. began to sizzle as an innovative company, causing its value to take off and becoming the most valuable company in the world in 2013. Figure 38.1 offers a historic look at Apple's stock price since September 1984. Note the stratospheric rise of the stock beginning around 2005, which has not stopped since (the temporary, recession-induced dip in 2008–2009 notwithstanding).

Apple Computer, Inc. was incorporated in 1977. Until 2005, the share price reflected an ordinary mature company, but the share price rocketed up in 2005 when it began to introduce successfully an unprecedented string of new products (iPod, iPhone, iPad, and so on). Estimated value of the company reached US$415 billion as of March 2013, becoming the world's largest company based on market value. Renewed emphasis on bold innovations by CEO and founder Steve Jobs changed a mature company into an innovative growth powerhouse; because of the innovations, millions of investors valued Apple Inc. very highly, as reflected in the stock price since 2005. Other examples of stock valuation and innovation discussed below are:

1. Amazon.com – continues to be innovative after maturity (see Figure 38.2).
2. Earthlink – lost its innovativeness (see Figure 38.3).
3. Yahoo! – lost its innovativeness after maturity (see Figure 38.4).
4. Taser – lost its innovativeness after maturity (see Figure 38.5).
5. Google – Innovative and growing (see Figure 38.6).

Example 1: Apple Inc.

Facts (Figure 38.1)

- In Figure 38.1, peak share price: $620 in April 2012, decades after it began to trade on the exchange; in the figure, share price was a few dollars in the eighties and nineties but exploded after 2005;
- Market value: $ 577.1 billion, July 2012;
- Total revenue: $ 148.8 billion, July 2012;

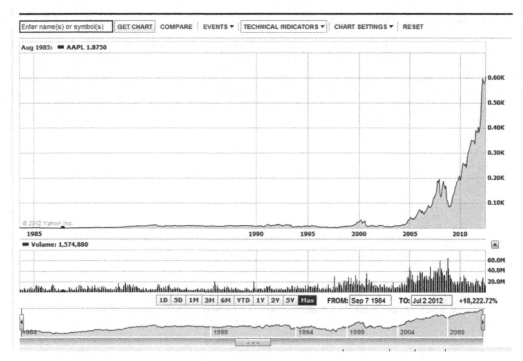

Figure 38.1. Apple stock price between September 7, 1984 and July 2, 2012

(*Source*: Yahoo! Finance)

- Profit margin: 26.97%; operating margin: 35.62%; note that other companies in this list do not compare well with Apple Inc. except Google.
- Debt: $0; cash: $27.65 billion;
- July 2012, share price, $607; 98 percent of peak; rapid growth in value since 2009.

Partial Explanation

Note that profit margin and operating margin for Apple Inc. are very high. Cash on hand is more than $27 billion, with no debt. Apple share price has exploded after the introduction of iPod, iTunes, iPhones, iPad, and their new generations, one following the other rapidly, may explain the share price behavior of this company; at $577 billion (August 3, 2012), it was the most valuable company in the world based on market value (i.e., share price times shares outstanding). The company has created an image of a pro-lific disruptive technological innovator, which contributes positively to the share price.

Question

What do you learn from Apple?

- New products that consumers want explains business success. This illustrates the theme of the book: *Engineers/scientists, devote your time to developing good products, the rest will follow.* This is merely one lesson.

Example 2: Amazon.com

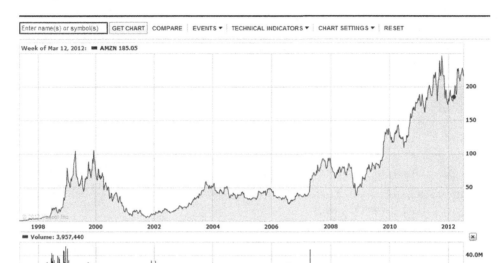

Figure 38.2. Amazon stock between May 16, 1997 and July 16, 2012

(Yahoo! Finance)

Facts (Figure 38.2)

- Peak share price about $246 in late 2011 in the figure;
- Market value: $106.2 billion, July 2012;
- Total revenue: $54.33 billion, July 2012;
- Profit margin: 0.69%; operating margin: 1.17%;
- Total debt: $0; cash: $4.97 billion;
- July 2012, share price $217; 88 percent of the peak eight months after the peak.

Partial explanation

Share price for Amazon has taken off, driven by the introduction of Kindle, Kindle Fire and numerous Internet retail innovations to expand target market and partners in Internet retail. This may explain this technology company's share price performance. Note that profit and operating margins are low compared to Google. Therefore, something other than profits is propelling the share price up; it could be the anticipation of innovation- and technology-driven growth.

Question

What do you learn from Amazon?

• Technology-driven growth is good for a business; engineers take note.

Example 3: Earthlink

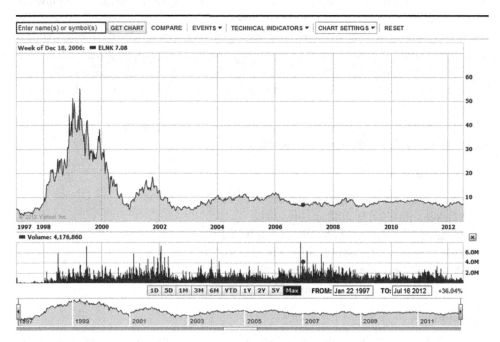

Figure 38.3. Earthlink stock between January 22, 1997 and July 16, 2012

(Yahoo! Finance)

A Note on Sky Dayton, Founder

From the age of nineteen in 1990, he invested in small coffee shops and other businesses. He describes how he got the idea for Earthlink (from his speech while receiving an award at the University of Southern California):

> And I heard about this thing called the Internet. I thought, that sounds kind of interesting. The first thing I did is I actually picked up the phone and dialed 411 [phone directory], and I said, "I'd like the number for the Internet, please." And the operator is like, "What?" I said, "Just search any company with the word Internet in the name." Blank. Nothing. I thought, "Wow, this is interesting. What is this thing anyway?"

The above admission by the founder of Earthlink shows a form of market research taking place in his mind during this conversation with the 411 phone-directory operator in 1993. He saw the opportunity; the phone-directory operator had never heard of the Internet, and there was not a single entry in the Yellow Pages for "Internet." He tried to connect to the Internet and found it very difficult. It motivated him to grab the opportunity (he stumbled into it) for an idea for a new business. Consequently, he started one of the earliest Internet Service Providers (ISP) in 1994 with the goal of giving easy access to the Internet for customers; this is a Phase 1 event in the life of Earthlink and Dayton, the innovator.

It was in 1994, at twenty-three, he decided to start Earthlink to provide Internet access to customers. Lacking investment capital, he approached a friend's father for angel investment and gave up 40% of the company for the first $100,000 investment. Dayton started in a 600 sq. feet office in Los Angeles and provided nationwide Internet access service by the summer of 1995.

It was December 1991 when the first Internet server appeared in the United States. In February 1993, the National Center for Supercomputing Applications (NCSA) at the University of Illinois, Urbana, released the first version of Mosaic, the first web browser. There were very few Internet service providers then. Dayton heard about the Internet in 1993, and jumped at the opportunity to sell access to the Internet; the rest is history, and the growth of the Internet is unparalleled. In twenty years, Internet-based trade has easily climbed over a trillion dollars!

Facts (Figure 38.3)

- Stock price for Earthlink peaks at $55 in 1999;
- Market value: $ 671 million, July 2012;
- Total sales revenue: $1.39 billion in July 2012;
- Profit margin: 1.28 %; operating margin: 8.61%;
- Debt: $ 652 million; cash: $257.9 million;
- July 2012, share price at about $7.20; 13.1 percent of peak.

Questions

1. Market value of the company is down substantially; are innovations keeping up with customer expectations and competition?
2. What is the effect of competitors on Earthlink?
3. Are there many new innovations in this industry?

This industry has low innovation and it is now a utility such as electrical power, water and gas suppliers.

Example 4: Yahoo!

Facts (Figure 38.4)

- Peak share price about $105 in late 1999;
- Total revenue: $4.98 billion, July 2012;
- Profit margin: 22.13%; operating margin: 15.91%;
- Market value: $19.5 billion, July 2012;
- Debt: $130 million; cash: 1.91 billion;
- July 2012, share value $15.60, about 15 percent of the peak.

Questions

1. Market value of the company is substantially down since the peak; are innovations keeping up with customer expectations and competition?
2. Google entered the market many years later as a competitor – see below. How is Google doing relative to Yahoo!? Why?

Google has done relatively well based on innovation-based growth strategy.

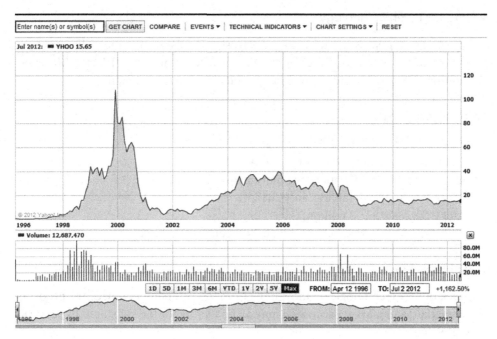

Figure 38.4. Yahoo! Finance stock between April 12, 1996 and July 2, 2012

(Yahoo! Finance)

Example 5: Taser

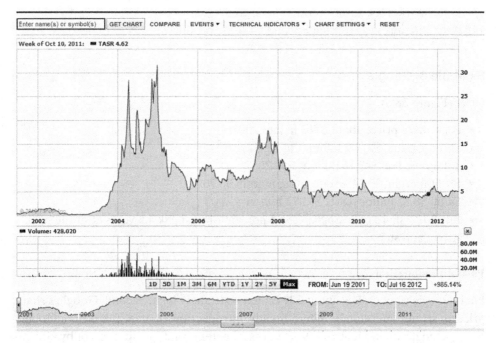

Figure 38.5. Taser stock from June 19, 2001 to July 16, 2012

(Yahoo! Finance)

A Note on Taser Technology

Taser's main product is a non-deadly electroshock weapon for the police/military. It uses electric current for "neuromuscular incapacitation" and "Electro-Muscular Disruption (EMD)," causing strong involuntary muscle contractions.

Facts (Figure 38.5)

- Peak share price about $32 in late 2004;
- Total sales revenue: $99.5 million; July 2012
- Profit margin: 2.49%; operating margin: 6.38%;
- Market value: $290 million, July 2012;
- Debt: $0; cash: $23.2 million;
- July 2012, share price 5.23, about 16.3 percent of the peak.

Partial explanation

The lack of new technological innovations from the company, competitors' products in the market, and market saturation may all be factors for the substantially reduced market value.

Questions

1. Market value of Taser is down substantially; are innovations keeping up with customer expectations and competition?
2. Are there new competitors in the market after the peaking of share value?
3. Is the market for Taser products saturated?
4. Are there any negative legal issues with the product?

Example 6: Google

Facts (Figure 38.6)

- Peak share price: about $710 in late 2007;
- Market value: $209.7 billion, July 2012;
- Total revenue: $43.16 billion, July 2012;
- Profit margin: 25.74%; operating margin 30.76%;
- Debt: $8.12 billion; cash: $41.72 billion;
- July 2012, share price $606.94, 85.5 percent of the peak 4.5 years after the peak – strong and stable.

Partial Explanation

First, Google's net cash is about $42 billion, extremely high. Second, Google's share price is strong and steady. Why? Steady introduction of new features – Google Earth, Google street view, ability to access all patents on Google, attempt by Google to place all library books on Google, fee-based services with each new capability,

Figure 38.6. Google stock from August 19, 2004 to July 16, 2012

(Yahoo! Finance)

self-driving cars, and so on – may partially explain the success of this technology company. Competition is unable to get ahead of Google on technology innovation. How does Google compare with Yahoo! on new technology introduction?

Questions

1. What do you learn from Google, the Phase 7 technological innovator?
2. What is unique about Google's range of products?
3. What could Yahoo! learn from Google?

Lessons from All Six Companies

a. Technological innovations took Apple stock to new peaks, from about $30 in 2005 to nearly $650 in 2012; from 1983 to 2005, the share price was not notable; recent growth is remarkable by any measure.
b. In recent years, without notable innovations valued by the market, Yahoo! Earthlink, and Taser are around 15 percent of their peak market values;
c. Google, with its bold innovative moves, maintains it market value close to the peak since late 2007 (except during the broad stock market crash during 2009).

Recognize that continued technological innovations contribute to the valuation of a technological company. Technology companies must sustain company market value with newer technologies. Such companies need significant investment in research and development. They must seek and employ a number of innovators. Today, some companies provide an environment for creative people to work with bold ideas and bring them to the market; Google, Apple, and Amazon are known to provide such environments to promote innovation.

Questions

You may need to conduct additional research to answer these questions.

A. Why did Google exceed Yahoo! in total sales and market value, although Google started much later with a product similar to Yahoo! at inception?
B. Considering all six companies, what are the lessons for technology companies that have reached maturity?

Comparing Phase 7 Firms from Different Industries: Apple and Walmart

Technological innovation can be very profitable for start-ups and established companies. A small percentage of all firms are technological innovators – Apple Inc. is a good example. Walmart Stores is a very successful retailer in a very different industry. A comparison of their financial statements from *Yahoo! Finance* reveals how successful firm in one industry (large-scale retail) is very different from a successful firm in another industry (personal smart electronics such as smartphones, tablets, laptops, etc.).

In Table 39.1, the last row was computed by dividing the net income by total revenue to find the net income per dollar of sales. Walmart makes 3.5 cents of net income from every dollar of sales, whereas Apple makes 23.94 cents per dollar of sales. Apple is 6.84 times more profitable than Walmart per dollar of sales; *that is the power of technological innovation–based businesses* that can generate considerably more profits per dollar of sales than other companies in staple industries. That is why, the valuation of a high-technology company may be five to ten times greater than a traditional business with equivalent sales in dollars.

Walmart is not an inferior company; it is the best or one of the best retailers in the world. But in the industry where Walmart competes (large discount retailers),

Table 39.1. *Walmart and Apple Financials*

	Walmart, Jan 30, 2012	*Apple, Sept 23, 2011*
Total revenue or sales	$446.950 billion (B)	$108.249 billion (B)
Total cost	$335,127 B	$64.432 B
Gross profit	$111.823 B	$43,818 B
Operating income	$26.558 B	$33.790 B
Income before tax	$24.398 B	$34.205 B
Net income (i.e., profits)	$15.669 B	$25.922 B
Market value	$252 B	$577 B
Net income per dollar of sales	**3.5 cents**	**23.94 cents**

Source: Income statements for both companies on July 30, 2012; http://finance.yahoo.com/q/is?s=WMT+Income+Statement&annual
http://finance.yahoo.com/q/is?s=AAPL+Income+Statement&annual

net profits per dollar of sales will always be low. Apple specializes in various niches of a technology industry where net profits per dollar of sales are expected to be very high. In this industry, quantum changes in technological innovations are possible and eagerly embraced by customers. Moreover, Apple has created an image of being the top innovator in a number of high-tech products including, among others, iPod, laptops, smartphones, and iPad. Such an image boosts Apple's net profits per dollar of sales.

Question: If you are the CEO of Apple, what goal would you give to your employees concerning net income per dollar of sales? Keep it where it is now or make it even better? If the latter, how can that be achieved?

PREPARING TO BE AN ENTREPRENEUR

40 Teams and Teamwork

Innovation inside small and large corporations needs good teamwork. We do not always have the chance to pick our team members, and even if we do, team members are challenged in their ability to work with some members of their team. Why? People differ from each other in many respects:

1. Personal motivation, inspiration, expectations, and the desire to please others;
2. Ability for relationship development with others;
3. Ability to work with details, or the ability to see the big picture;
4. Ability to get things done within a time limit, remain focused on goals, and make quick decisions;
5. Ability to tell others what to do, or give advice;
6. Ability to make logical decisions, and the need for extensive data before making a decision.

The differences among people we work with are partially the result of differences in personality styles. Learning to understand and work with diverse personalities is critical to a manager or team leader.

DISC: Four Personality Styles

One popular training model called the DISC model, originally introduced in the 1920s by psychologist William Marston, says that all people fall under four different personality types. This model has survived the test of time. It is a very simple yet powerful tool for teaching personality types to leaders and team members for working with others more effectively; the model also has an uncanny way of improving our understanding of self and others.

The four personality types are Dominant [**D**], Influence seeking [**I**], Steady or peace-seeking [**S**], and Conscientious [**C**].[1] There are considerable differences among the four personality styles.

---

[1] A full explanation of the different personality types can be found at www.changingminds.org

To cement the idea of personality types in your mind, associate the personality type with examples: Steve Jobs and Donald Trump are strong **D** personalities, always eager to change or move something; you may know of successful sales professionals with strong **I** personality; sports coaches such as John Wooden (at UCLA, won ten NCAA basketball championships in twelve years, seven in a row), who was able to understand, analyze, and communicate the minutiae of basketball to his players, are strong examples of a **C** personality; and you may know examples of **S** personalities, who are not out to change the world.

Personality Broadens with Age

While personalities may seem to be cast in stone early in life, with age, people become more skilled at additional personality styles that are not natural to them. For example, a teenager may be very high in **C**, but ten to twenty years later, as a professional, may have developed the **I** dimension to become a more balanced individual and a more effective employee and family member. It is not uncommon for a person to have one primary, strong personality type with a secondary personality type that is also noticeable to others.

How to Build Better Teams?

With time, a leader or team member can become reasonably good at ascertaining the personality styles of people with whom they work. With an understanding of personality styles from a system like DISC, one can substantially reduce the chances for misunderstanding, miscommunication, and conflict, and thereby experience better teamwork and effectiveness.

A leader can make working relationships most effective by accommodating individual strengths and weaknesses and by fitting the task to an employee's strength. For jobs requiring a strong, decisive person, who is not likely to be pushed around by people or circumstances, a **D** personality would be a better fit for certain professions. For a job requiring considerable people contact and people skills, such as sales, an **I** personality would be a better fit. For a job needing a lot of attention to detail and minimal supervision, such as computer professionals and accountants, a **C** personality may be a better fit. For a job needing a lot of people contact without being overbearing or threatening, an **S** personality may be better suited for the job. Do not demand something an individual's personality style cannot deliver. For example, do not send a difficult message to deliver to a **D** personality through an **S** personality. More than once, an S personality, when asked to deliver a difficult message to a D personality, has responded politely, "It would be better if someone else did it."

Other Personality Models

Another model used by those who train corporate teams for innovation is the FourSight model.[2] It recommends innovation teams composed of members with varying innovative skills found in the Ideator, Clarifier, Developer, and Implementer. An Ideator originates new ideas for innovation, a Clarifier can define the problem and explain the product and its capabilities to different audiences, and a Developer can develop and improve the product to better address the problem, better satisfy customers or better address competition. Finally, an Implementer can take the idea and product to the market to enable commercial success. In rare occasions, a single individual is endowed with all these skills; Steve Jobs of Apple and Jeff Bezos of Amazon are examples of rare individuals with multiple skills. Even they still needed team members to fill the void in their skills or in the process of scaling up their business.

A diverse team is an asset. By promoting and encouraging multiple strengths that the personality styles of team members and their innovative strengths bring to a team, the team can accomplish more and avoid destructive bickering.[3]

However, some innovative individuals do not perform well in teams. Forcing them into a team will be counterproductive.

[2] www.foursightonline.com/about.html
[3] A version of this chapter appeared in American Management Association's *AMA Shift* as a guest blog under the title, "Building teams means understanding strengths and styles," June 13, 2012, www.amanet.org/shift/index.php/2012/06/13/building-teams-means-understanding-strengths-and-styles (accessed July 14, 2012).

41 Leadership Issues in Start-Up Businesses

Jay Clark

Engineers and business leaders may deal with processes, products, and services with relative ease, but they may face unexpected challenges dealing with different stakeholder groups such as investors, managers, customers, and employees. In a newly forming business, many of the relationships are in a flux and evolving. For this reason, soft skills needed to be an effective entrepreneurial leader in a newly forming business is a greater challenge compared to leadership challenges in a stable ongoing business.

Most literature on starting a new venture focuses on the tasks such as business plan, product sourcing, financing, marketing, and so forth. What may be lacking is the attention to soft skills needed to lead the start-up to success. If leaders do not pick up leadership skills along the way, their success may be limited by their own destructive or inadequate leadership.

The most common destructive leadership traits result in either questionable or unethical decisions. Destructive leadership is mostly short-term biased. Many destructive forms of leadership may be on display during the start-up process when a leader must cultivate healthy behavior, actions, and culture in the new organization. The end result of the destructive leadership during the start-up phase is often a failed start-up. How does one detect some common destructive leadership behavior?

Leadership Behavior

When leaders demonstrate destructive leadership, they lose sponsors, supporters, followers, employees, investors, partners and others vital to the success of the start-up. According to experts, there are three types of destructive behavior in some leaders: egocentric behavior (narcissism), lack of self-awareness, and a refusal to make personal sacrifice. The opposite of these are: humility, the ability to

Jay Clark, PhD, MBA, is a leadership consultant and founder of Nextgen Global Leaders Program.

understand and interpret own emotions and purpose, and the willingness to make personal sacrifice.

Humility

Egocentrism damages or limits the ability of the leader and his/her organization to learn and adapt to change (Morgan, 2006). Egocentrism is the opposite of humility, which downplays or keeps in check the relative importance of self. A destructive ego is evident when we "get a distorted image of our own importance and see ourselves as the center of the universe" (Blanchard, 2007, p. 267). Such leaders may appear to be vain and arrogant (Roberts and Wood, 2007), with consequent negative effect on the organization.

What is an example of "vain" conduct? During a time of financial hardship for the start-up, a vain leader may seek an expensively designed and furnished CEO's office, and/or unaffordable automobiles at company expense; such misplaced priority may harm the financial health of the start-up. Vain conduct compromises the long-term purpose and success of the organization – that is not leadership. In contrast, at Amazon.com, the principals were improvising desks out of old doors as late as four years into the company's existence while waiting for profits (see Chapter 36). That is a mark of true leadership.

Self-Awareness

Self-awareness is the ability to understand one's own emotions while being clear about one's purpose. When leaders have control over their emotions and have better self-awareness, they are more adaptable to changing environments (new start-ups are faced with changes daily) and are successful in overcoming obstacles. Not only do successful leaders understand their own emotions; they are also able to effectively communicate their passion in the context of the overall purpose of the organization; they are endowed with valuable intuition in addition to technical expertise and common sense.

If a leader lacks self-awareness, he or she is likely to succumb to knee-jerk reactions during moments of disappointment, surprise, adversity, conflict and tension. These moments give rise to anger, disappointment, fear and other negative emotions. A self-aware leader would not let them take over or dominate his/her decisions. A good leader would deal with disappointment, surprise and adversity without compromising the ultimate purpose of the organization.

To use an example from tennis, if one of the doubles' partners makes a mistake or error in judgment (happens often during the course of a match) and causes the other partner to react with uncontrollable negative emotions, the team result would certainly suffer, and the team would break up eventually.

Self-Sacrifice

The self-sacrifice of leaders demonstrates commitment of the leader to the group or cause, develops stronger followers and increases the performance of followers in the group.

Findings by van Knippenberg et al. (2005) provide empirical evidence to support the fact that self-sacrifice has a positive effect on performance. Self-sacrifice of the leader is evidence to others that the leader is dedicated to the group. Self-sacrifice from a leader also allows the followers to see more clearly the long-term purpose of the organization. Finally, it enhances the group's motivation to follow the leader; an employee more engaged with the leader increases the performance of the organization and its success.

Self-sacrifice of the leader for the cause of the start-up builds credibility among the employees. The leader's contributions toward developing or creating products, willingness to invest in the company, borrow and take risks also build credibility, as well as the leader's prior experience and competence.

Concluding Note

Before a leader can influence others, they have to learn to lead themselves; a self-aware leader does that.

42 What We Know about Entrepreneurs

(Anonymous Contribution)

While some people may believe entrepreneurs are born, researchers have shown that many aspects of entrepreneurship can also be taught and learned (Morrisette and Schraeder, 2007). Entrepreneurs are called to perform a complex, multifaceted task that cannot be easily defined; training could help. In this short chapter a summary of research on entrepreneurs is provided for new entrepreneurs. Based on research literature, the characteristics that are associated with entrepreneurs are:

1. Entrepreneurs voluntarily *accept the challenge* of starting and owning a business, and may be relentless in their efforts to expand their business (Peel & Inkson, 2004).
2. Entrepreneurs go beyond thinking and reasoning; *they act* in pursuit of ideas that they perceive to have merit or represent worthwhile opportunities (Morrisette & Schraeder, 2007).
3. Entrepreneurs are *good at experimentation* that enables successful start-up businesses; experimentation is the process of testing a with data to discover unknown or hidden truths before taking action (Hölzl, 2010).
4. Entrepreneurs often face and *overcome resistance to their ideas.*
5. Entrepreneurs *take measured and calculated risk*; business start-ups face many risks. "Risk" refers to the fact that the desired success of an entrepreneurial venture is not certain – that is, the chance of the venture's success is much less than 100 percent.
6. Entrepreneurs *devote distinctive energy and drive* to the entrepreneurial cause.
7. Entrepreneurs often *commit their own resources* in the form of time and money to further the entrepreneurial cause initially.

Those who are less willing to take significant risks are known as conservative risk-takers or risk-averse people. Such entrepreneurs may prefer to minimize perceived risks by committing their energy and resources only to ventures that are not perceived to be too risky – that is, those for which the probability of success appears to be better than other options. In contrast, risk-tolerant entrepreneurs are those who tolerate a higher level of risk in the new venture, either knowingly or unknowingly.

The entrepreneur's level of expertise or experience associated with a particular product/service at the core of the business venture may help reduce the risks facing the entrepreneur. Relatively speaking, an entrepreneur tolerates or accepts more risks than an average person does.

Regardless, before starting a venture, an entrepreneur must carefully evaluate the risks and potential success of the new venture in the context of his or her knowledge, expertise and experience. Starting a business venture as a team could bring added knowledge, experience and expertise to the venture, thereby increasing the potential for success and reducing the risk of failure. However, given that the early years of a business are challenging and sometimes chaotic, the ability of the team to survive under stress will be critical to the success of the new venture.

Creating Value as an Engineer in India

Hephzibah Stephen

I Created Value for My Employer

After completing my engineering degree, I joined Hindustan Teleprinters Limited (HTL), a fully owned subsidiary of the Government of India. As a young engineer in R&D, I developed the idea of using the Public Switched Telephone Network (PSTN) to send telegrams. This option was cheaper for the Indian government's Department of Telecom munication (DOT) than maintaining the telegraph network through specially trained operators.

I designed and developed a "Rural Messaging Terminal" (RMT), which transmitted Morse code (Telegrams) over the PSTN telephone line. The prototype was approved by the Telecom Research Center and cleared for production. It was manufactured by my company (HTL) and sold to the Government of India, which deployed them in the DOT network across India. Since all telecommunication services and network assets were owned by the Government of India at that time, they were also the only customer for our product. Overall, my contribution for the RMT program spanned design, prototyping and obtaining the technical-engineering clearances required for our products' use in the government DOT network.

Around 1980, the product that I was associated with was very successful for HTL; it was a capital expenditure item for our customer. We manufactured and sold

This chapter has been included because it is relevant to engineers in countries such as India, China, and other rapidly developing nations. In India, both the government and private entities are committed to creating a strong and widespread start-up ecosystem. Indianstartups.com and other private entities are providing valuable service to the inventing and start-up community. Government of India's Department of Industrial Policy and Promotion has announced a new "Start up, Stand up India" program to invigorate the entrepreneurial ecosystem in India; as part of this new program, many incentives are expected, including tax breaks for angel inventors. International firms operating in India are also credited with promoting, supporting and inspiring the growth and development of innovation and a vigorous start-up ecosystem in India; Google India has launched a highly visible start-up contest in India. It also mentors start-ups in early and advanced stages in India, and has a full-time staff promoting the start-up ecosystem in the country. Google India is just one of many international and national companies boosting innovation and the start-up ecosystem in India. One may find similar efforts in other countries, such as China.

several units over the next three years in India, contributing to significant annual revenues for the company, which also helped maintain employment on the manufacturing side of our business. Through the RMT program, I created value and profit for HTL, its employees, the Government of India and the public served by the equipment.

I Also Created Value as an Entrepreneur

It was my dream to be an entrepreneur. Therefore, after twenty years at HTL, when an opportunity presented itself in 1998, I stepped out on my own.

During the late 1990s, India was fast becoming a popular destination for a variety of offshoring and outsourcing businesses for U.S.-based companies. American entrepreneurs, with investment dollars, were looking for local partners for new start-up operations in India.

The first venture I investigated was with a business partner from the United States. I spent two months in California learning about his business. However, upon returning to India, I realized this was not a viable business for me. For a time thereafter, I did freelance work as a technical writer, writing user manuals for a local software products company.

I then met an American businessman with companies in various emerging markets and looking for an opportunity to expand into India. We established XYZ Datasoft Private Limited, a captive subsidiary of his company in the United States. Based on market analysis, we started medical transcription services for a client in the United States, who provided medical transcription services to hospitals and doctors in the United States. All marketing and customer relationship activities were the responsibility of the U.S.-based parent company.

Based on various benefits and incentives offered by the Indian government for start-ups in the information technology (IT) sector, I set up Datasoft as a 100 percent Export Oriented Unit (EOU). In order to receive government approval, I submitted a detailed business plan, cash flow and profit and loss (P&L) analyses to enjoy benefits such as the use of 100% foreign equity (needed government permission), exemption of excise duty, reimbursement of Central Sales Tax, and also exemption from corporate income tax for a period of time.

Upon completion of the necessary paperwork and approvals, my company devoted time to organize an IT infrastructure needed to deliver the transcription services as committed to our customer. Once initial operations were under way, staffing projections for the next five years based on our business plan were developed, and we proceeded to identify and hire the right people. After four years of successful operations, we also diversified into the lucrative software development and application maintenance businesses.

I managed Datasoft for a total of eight years. My responsibilities during this time were to flexibly manage company operations to meet changing market requirements, ensure quality and reliability and deliver value to the investors. While doing all this, I also had the satisfaction of providing a high-income, high-quality-of-life work environment for the hundreds of employees who passed through our doors. At our peak, Datasoft had a workforce of more than eighty people, including people with severe physical disabilities. Currently, I serve as a board member; I have not been involved in day-to-day operations since 2008. This company has created value for employees and investors for many years.

Lessons for Other Engineer-Entrepreneurs

Typical engineers have both advantages and disadvantages when being entrepreneurs. Engineers have the advantage of education and training in college on cutting-edge technology, which provides them the opportunity to see market opportunities for evolving technologies before they become widely known. Additionally, engineers have a natural inclination toward analytical problem solving; when it is combined with the ability to create a "product," engineers can add considerable value to new business start-ups.

As an entrepreneur, I learned that an engineer-entrepreneur must have a new-product-creating "prototype" mindset as well as a "production" mindset, where the product is modified to address customer needs and business constraints such as cost targets, manufacturability, marketability and others.

I learned that hiring the right person for the right job and retaining good employees is very important for a successful business enterprise. There is also a constant need for managing relationships and expectations with employees, suppliers, customers, investors and other business partners. This makes people management and negotiation skills crucial to the success of the business.

As an entrepreneur, I gained special appreciation for the business functions such as marketing, sales and distribution. Knowledge and skills in these areas can be either developed or acquired through suitable partnerships. As an engineer-entrepreneur, if you think you do not have the time or the best ability to deal with business and people side of your business, find a mentor or a partner. I benefited from working with a partner experienced in marketing and business development; I would recommend the same to any new engineer-entrepreneur.

44 Technology Licensing Option for Inventors

Some inventors may not want to become entrepreneurs but continue to be serial inventors devoting their time entirely to more inventions rather than learning the art of starting as well as running a new start-up business. Such inventors are encouraged to consider licensing their inventions after filing a valid patent application with the USPTO. This option, if executed, could take away the need to start a new business to create income for the inventor; after all, starting a new business with one's invention is to create an income. Licensing could produce a steady income to the inventor through royalty agreements, while allowing the inventor to continue to invent without interruptions. A sound royalty agreement between the inventor and the licensee could be prepared with the assistance of legal professionals, while the cost of the associated expenses could be assigned to the licensee or to future cash flows to the inventor, or both.

Licensing an invention could occur after either a Provisional patent application is filed, or after a Nonprovisional Utility application is filed. If a Nonprovisional application is not filed before the licensing application, the licensee may be required to file such an application as part of the agreement. This may be a good option for the inventor if the cost of Nonprovisional patenting for a complex invention is prohibitive. Further, international patenting in other nations could be too expensive for pro se inventor-applicants to afford. One of the benefits of licensing is the ability of individual inventors to require the licensee to seek international patent rights as part of the licensing agreement.

Inventors interested in licensing must seek out law firms that specialize in licensing agreements between an individual inventor and an interested business. Organizations such as the United Inventors Association of America can help identify law firms with a good record of working with the individual inventor. There are also helpful books for inventors on licensing new ideas and technologies.[1]

---

[1] E.g., Reece, H., *How to License Your Million Dollar Idea,* 2nd edition, New York: John Wiley and Sons, 2002.

APPENDIXES

THE LANGUAGE OF ENGINEERS

Units, Specifications, Tolerances, and Drawings

In communicating with businesses and non-engineers, engineers must know and use engineering-specific languages, which are units, specifications, tolerances, and drawings.

Units

Everyone is aware of the "horsepower" of a car; it is a very specific unit of power measurement devised by engineers (see below for various units of power).

There are at least three types of units that US engineers come across. The 1824 Act of the British Parliament established the more rudimentary Imperial units such as the yard, pound, bushel and gallon. Later US standards established in 1834 were named after the British units, but they were not the same; the gallon is not the same in the two different systems of units.

Today, the most widely used system around the world is the Systeme International d'Unites (SI), which is also called the Metric System; it originated in France in 1790.

Engineers need extensive sets of units to measure both tangible and abstract engineering entities; there are literally hundreds of units to measure different engineering entities. The following is a listing of common engineering units.[1] The listing in Table A.1 provides both metric and nonmetric units as well as the conversion between the two.

---

[1] For more, see http://www.csgnetwork.com/converttable.html or www.rapidtables.com, as well as http://www.efunda.com/units/show_units.cfm?Alfa=A&String1=aaaa&String2=azzz

Table A.1. *A small sample of more common units*

Common base SI units

1.	Length	metre (meter)	= m
2.	Temperature	Celsius	= C
3.	Energy	joule	= J
4.	Power	watt	= W
5.	Electric current	ampere	= A

Derived SI units

1.	Area	square meter	= m^2
2.	Speed	meter / second	= m/s
3.	Acceleration	meter / second squared	= m/s^2

English engineering units vs. SI units

1.	Time:	second (sec)	vs. second (s)
2.	Length:	foot (ft)	vs. metre (meter) (m)
3.	Mass:	pound mass (lbm)	vs. kilogram (kg)
4.	Force:	pound force (lbf)	vs. newton (N)
5.	Temperature:	Fahrenheit (F)	vs. Celsius (C)

Time

1.	Seconds (sec)	= 1s
2.	Minute (min)	= 60s
3.	Hour (hr)	= 3,600s
4.	Hertz (Hz);	1 Hz means one cycle (repetition) per second

Distance/length

5.	Foot (ft)	= 0.3048m
6.	Mile (mile)	= 5,280 ft
7.	Yard (yd)	= 3 ft
8.	1 m = 1000 mm = 3.281 ft	= 39.37 in
9.	1km = 1000m = 0.621 miles	= 1,094 yards

Mass

10.	Gram (g)	= 0.001 kg
11.	Pound mass (lbm)	= 0.45359237 kg
12.	Carat (metric)	= 0.2 g
13.	Short ton (ton)	= 2,000 lbm
14.	Long ton (ton l)	= 2,240 lbm
15.	Metric tonne	= 1,000 kg = 2,205 lbm
16.	Ounce (oz)	= 28.34952 g

Gravity

17.	Gravity's acceleration (G)	= 9.80665 m/s^2

Force or weight (SI)

18.	Newton (N)	= 1kg m/s^2

Energy

19.	Joule (J)	= Nm
20.	British thermal unit (BTU)	= 1,055.056 J
21.	calorie (cal)	= 4.1868 J (international)
22.	Calorie (Cal)	= 4.1868 kJ (nutritionist's)

Power

23.	Watt (W)	= J/s
24.	Horse power (hp)	= 550 ft lb/s

25.	1 hp (mechanical)	= 745.669 W
26.	1 hp (electrical)	= 746 W
27.	1 hp (metric)	= 735.49 W

Electrical SI units

28.	Ampere (A)	= W/V = electric current
29.	Watt (W)	= V x A = power
30.	Volt (V)	= W/A = electric potential
31.	Ohm	= V/A = electric resistance

Example of motorcycle battery specifications

1. 12 Volt
2. 18 Amp hour (Ah)
3. 270 Cold cranking Amps (CCA)
4. Nut + bolt terminal
5. Dimensions 6.9" × 3.5" and × 6" (USA)
6. Weight 12.3 lbs (USA)

The above is only a small sample of units under each heading; there are hundreds more.

Drawings

Engineering drawings are meant for design engineers to communicate to various users of the drawing. Drawings communicate what the product looks like, its dimensions and tolerances for the manufacturing department, suppliers, inspectors and approval-granting government agencies. Furthermore, drawings communicate essential engineering information to subcontractors, who may bid on the product to enter into a binding contract to make and supply the product in the drawings; the bid is tied to the drawings that invariably include dimensions, specifications and tolerances.

Patent applications require drawings, but they may not have all the information contained in a drawing prepared for communicating to manufacturers. Drawings meant for manufacturing must include dimensions and tolerances.

Tolerances

In addition to dimensions shown in drawings, engineers will add tolerances to design dimensions to communicate to the manufacturer the extent of deviation from the designed dimension allowed during manufacture without compromising the performance of the product. During inspection of the finished product, inspectors would check to see if the dimensions of the manufactured product are within the tolerance limits provided by the design engineer.

Tolerances are the acknowledgment by engineers that to make something to the *exact* specifications/dimensions specified by the design engineer is sometimes too expensive and unnecessary. For example, in the case in Table A.2, head area is specified as 120 sq. in. for one of the tennis racquets. Let us assume for discussion's

Table A.2. Two examples of specifications – Wilson Tennis rackets

Specifications	Cierzo 120	Pro Staff 6.1
List Price, July 2012	$396	$293
Head area (sq. in.)	120 (more material)	90
Length (in.)	27.9	27
Beam width[*] (mm)	29 mm dual taper (uses more material)	17 mm flat
Composite (Basalt is unique to Wilson)	Basalt, Karophite Black	Basalt
String pattern	16 horizontal strings 19 vertical strings	16 19
Balance (location of center of gravity)	14.60 head-heavy	12.0 head-light (opposite effect)
Best-suited swing	Slow and compact (Target market: aging club players)	Longer, faster strokes (Target market: professional)
Prominent users	Local recreational players	Federer, one of the great professional players

[*] Beam thickness of the head is what you see in the side view of the racket.

sake that a design engineer specified tolerances for the head area as + or – 0.1 sq. in. During manufacturing, if the finished racquet has a dimension of 119.9 to 120.1 sq. in., it should be considered within the tolerance and acceptable for sale. Similarly, if the product was outsourced (i.e., purchased from another company), Wilson Company would accept the outsourced racquet if the head size is between 119.9 to 120.1 sq. in. Thus, tolerances are a useful and practical way of communicating between the designer and manufacturer, and a method for controlling cost and quality without any deterioration in designed performance.

If the tolerance is "tightened" to + or – 0.01 sq. in., the cost of manufacturing the tennis racquet would most likely increase. If no tolerances were allowed by the design engineer (no deviation from 120 sq. in. to be accepted), the cost of manufacturing could be prohibitive because all manufactured racquets that are not exactly 120 sq. in. must be discarded.

Consistent Use of a Unit System

Table A.2 is based on data provided in an advertisement for tennis racquets in the United States shown in Figures App 1.1 and App 1.2 in the United States; it has specifications for the tennis racquets in inches (head area in square inches) and beam thickness in millimeters (mm) from two distinct unit systems. This is rarely done in industrial practice. It is important to use the units conforming to one system in a drawings and communications between the designer and manufacturer to prevent misunderstanding or misreading of the units.

Figure App1.1. Wilson Cierzo 120 Blx; year 2012

Figure App1.2. Wilson Pro Staff 6.1 90 Blx; year 2012

Appendix B

Illustrative Document: Non-Disclosure Agreement

SECRECY/NONDISCLOSURE AGREEMENT

Made as of the ___ th day of _____20__ (the "Effective Date"), by and between _____ _____("COMPANY") with offices at _____ _____, and Auburn University ("AUBURN"), with offices at Office of Technology Transfer, 570 Devall Drive, Suite 102, Auburn, Alabama 36832.

Whereas, AUBURN has developed technology related to _____ _____ ("Technology") and COMPANY wishes to evaluate the same for possible collaboration and/or commercialization; and

Whereas, the parties recognize that in order to accomplish this purpose, it may be necessary or appropriate for one party to disclose to the other, information or technology which is considered by the disclosing party to constitute its confidential or proprietary business secrets which include but are not limited to Technology.

Now, therefore, in consideration of the premises and to induce the disclosure of confidential information, the parties agree as follows:

1. The receiving party shall maintain for a period of three (3) years from the end-date of this Agreement the confidentiality of information disclosed to it or otherwise learned by it during or as a result of research, hereinafter "Confidential Information", provided such information is in writing or other tangible form and clearly marked as proprietary when disclosed, or is so designated in writing within thirty (30) days of such disclosure. In addition, any information provided to the Recipient by or on behalf of Discloser that, by its nature and content, would be readily recognized by a reasonable person to be confidential or proprietary to the Discloser shall also be deemed Confidential Information of the Discloser. This term for confidentiality shall survive any termination of this Agreement. The receiving party shall use the same level of care to prevent the use or disclosure of the Confidential Information as it exercises in protecting its own information of similar nature.

2. Confidential Information shall be used by the receiving party solely for the purposes of facilitating and conducting the research project. The receiving party shall not use said Confidential Information for the benefit of the receiving party or for the benefit of other parties without the consent in writing of the disclosing party. The receiving party shall not perform, nor have performed, any tests or measurements on Confidential Information disclosed in the form of technology samples for the purpose of determining such samples' method(s) of manufacturing.

3. Confidential Information disclosed by one party to the other shall remain the property of the disclosing party, and shall be returned upon written request. The receiving party may, however, retain one copy of the Confidential Information in its legal files for the purpose of determining its obligations hereunder, except in the specific instance of technology samples which shall be returned to the disclosing party upon request.

4. The disclosure of Confidential Information by one party to the other shall not constitute a grant by the disclosing party to the receiving party of any species of right, title, interest, or property in or to Confidential Information. No license or other right under any U.S. or foreign patent, copyright, or know-how is granted or implied by this Agreement.

5. The receiving party shall use reasonable efforts to disclose Confidential Information received by it only to those of its employees who must be so informed to enable that party to accomplish the purposes stated herein and who have been provided a copy of this fully executed Agreement and are bound by that party's obligations of confidentiality hereunder.

6. The receiving party's obligations of confidentiality hereunder shall not apply to the following:

   a. information which is now or hereafter becomes a part of the public domain;

   b. information known to the receiving party before disclosure to it by the disclosing party hereunder as evidenced by its records;

   c. information given to the receiving party by a third party having a right to disclose the same; or

   d. information which is hereinafter independently developed by the receiving party without reference to or benefit from the Confidential Information received from the disclosing party; or

   e. information which the receiving party is compelled to publicly disclose by judicial or administrative process, or by other mandatory requirements of law.

7. No amendment or modification of this Agreement shall be valid or binding on the parties unless made in writing and signed on behalf of each of the parties by their respective duly authorized officers or representatives.

8. Neither party shall have the right to assign or otherwise transfer any right or interest herein to any other person, firm, corporation or association without the prior written consent of the other party.

9. This Agreement shall be construed, interpreted and governed in accordance with the laws of the State of Alabama, notwithstanding the residence or principal place of business of any party, the place where this Agreement may be executed by any party or the provisions of any jurisdiction's conflict-of-laws principles.

10. The term of this Agreement shall be one (1) year from the Effective Date.

11. The above constitutes the full and complete Agreement in this matter by and between the parties hereto.

_____   AUBURN UNIVERSITY

By _____   By_____
Name: _____
Title: _____   Associate Provost and Vice President for Research

Date: _____   Date: _____

This document is illustrative and it may have undergone updates in recent years.

For the most current version, go to: https://cws.auburn.edu/shared/files?id=159&filename=blank-nda_current.pdf

Appendix C

US Patent Office Documents

Appendix C1

USPTO: Steps for Patent Application

The following is the listing of the process to submit a Provisional patent application (priority-date rights for 365 days) or a Nonprovisional or Utility patent application (Nonprovisional patents are also called Utility patents, once granted, the inventor gets patent rights for 20 years in the US) to the US Patent and Trademark Office (USPTO). This book draws the attention of the inventor to Steps 7a, 6a, and 4a in the list below. The chapters of this book are geared toward assistance to the pro se applicant (an inventor applying for patent without the services of an attorney) preparing a provisional or a Nonprovisional application.

First, such an applicant must apply to the USPTO for a Customer Number (Step 7b below; see Appendix C3 for application form). The process may take about three weeks to complete. The USPTO will send you the customer number and would require you to respond to them before issuing you a "Certificate" to access their Electronic Filing System (EFS) for filing patents online. Filing patents online is cheaper than filing on paper; paper-filing for patents may be discontinued altogether.

Next, the applicant may draft a provisional application using any granted/ issued patent published on the USPTO website or accessible on "GOOGLE Patents Search" as a model; it is recommended that the applicant use as a model one or more Utility patents issued by the patent office for a product that is similar or in the same product space. For example, if you are writing the patent application for a bread toaster, look through granted Utility patents for bread toasters or something that could be used for toasting. Use several granted Utility patents prepared by a patent law firm as a model for your patent application; patent attorney or the patent attorney firm is identified on the cover page of issued patents; see example in Chapter 13 and its Supplement 2.

The next Appendix (C2) gives more details about the Provisional patent application process. After you have drafted the Provisional patent application, use Step 4a to complete the application. You can contact by phone the EFS support services of the USPTO and the Patent Assistance Center (PAC) of the USPTO. Several of the author's students have used these services over the phone and report excellent help during the process to make a Provisional patent application.

Patent Process Overview (reproduced from the USPTO website; many items in the list are hot-linked to the USPTO website for additional information)

1. Step 1, Applicant – Has your invention already been patented?
   a. Search the Patent Full-Text and Full-Page Image Databases
      i. If already patented, end of process
      ii. If not already patented, continue to Step 2
2. Step 2, Applicant – What type of Application are you filing?
   a. Design Patent (ornamental characteristics)
   b. Plant Patent (new variety of asexually reproduced plant)
   c. Utility Patent (most common) (useful process, machine, article of manufacture, composition of matter)
3. Step 3, Applicant – Determine Filing Strategy
   a. File Globally?
      i. Need international protection?
   b. File in U.S.? – continue to Step 4
4. Step 4, Applicant – Which type of Utility Patent Application to file?
   a. Provisional or
   b. Nonprovisional
5. Step 5, Applicant – Consider expedited examination
   a. Prioritized Examination
   b. Accelerated Examination Program
   c. First Action Interview
   d. Patent Prosecution Highway
6. Step 6, Applicant – Who Should File?
   a. File yourself (Pro Se; after studying the process)
   b. Use a Registered Attorney or Agent (Recommended)
7. Step 7, Applicant – Prepare for electronic filing
   a. Determine Application processing fees
   b. Apply for a Customer Number and Digital Certificate
8. Step 8, Applicant – Apply for Patent using Electronic Filing System as a Registered e-Filer (Recommended)
   a. About EFS Web
9. Step 9, USPTO – USPTO examines application
   a. Check Application Status
   b. Allowed?
      i. Yes, go to Step 12
      ii. No, continue to Step 10
10. Step 10, Applicant – Applicant files replies, requests for reconsideration, and appeals as necessary
11. Step 11, USPTO – If objections and rejection of the examiner are overcome, USPTO sends Notice of Allowance and Fee(s) due

12. Step 12, Applicant – Applicant pays the issue fee and the publication fee
    a. USPTO Grants Patent
13. Step 13, Applicant – <u>Maintenance fees due 3 1/2, 7 1/2, and 11 1/2 years after patent grant</u>
14. <u>Download the Utility Patent Application Guide</u>

Appendix C2

USPTO: Provisional Patent Application

The following is an overview of the USPTO Provisional application process.[2]

Background

Since June 8, 1995, the United States Patent and Trademark Office (USPTO) has offered inventors the option of filing a Provisional application for a patent which was designed to provide a lower-cost first patent filing in the United States and to give US applicants parity with foreign applicants under the GATT Uruguay Round Agreements.

A Provisional application for patent (provisional application) is a US national application filed in the USPTO under 35 U.S.C. §111(b). A Provisional application is *not required to have a formal patent claim or an oath or declaration* (emphasis added). Provisional applications also should not include any information disclosure (prior art) statement since Provisional applications are not examined. A provisional application provides the means to establish an early effective filing date in a later filed Nonprovisional patent application filed under 35 U.S.C. §111(a). It also allows the term "Patent Pending" to be applied in connection with the description of the invention.

A Provisional application for patent has a pendency lasting 12 months from the date the Provisional application is filed. **The 12-month pendency period cannot be extended.** Therefore, an applicant who files a Provisional application must file a corresponding Nonprovisional application for patent (Nonprovisional application) during the 12-month pendency period of the Provisional application in order to benefit from the earlier filing of the Provisional application. However, a Nonprovisional application that was filed more than 12 months after the filing date of the Provisional application, but within 14 months after the filing date of the Provisional application, may have the benefit of the Provisional application restored by filing a grantable petition (including a statement that the delay in filing the Nonprovisional application was unintentional and the required petition fee) to restore the benefit under 37 CFR 1.78.

---

[2] Reproduced from USPTO.govhttp://www.uspto.gov/patents/resources/types/provapp.jsp# (accessed August 2014).

In accordance with 35 U.S.C. §119(e), the corresponding Nonprovisional application must contain or be amended to contain a specific reference to the Provisional application. For Nonprovisional applications filed on or after September 16, 2012, the specific reference must be included in an Application Data Sheet. Further, a claim under 35 U.S.C. §119(e) for the benefit of a prior Provisional application must be filed during the pendency of the Nonprovisional application, and within four months of the Nonprovisional application filing date or within sixteen months of the Provisional application filing date (whichever is later). See 37 CFR 1.78.

Once a Provisional application is filed, an alternative to filing a corresponding Nonprovisional application is to convert the Provisional application to a Nonprovisional application by filing a grantable petition under 37 C.F.R. 1.53(c)(3) requesting such a conversion within 12 months of the provisional application filing date.

Converting a Provisional application into a Nonprovisional application (versus filing a Nonprovisional application claiming the benefit of the Provisional application) will have a negative impact on patent term. The term of a patent issuing from a Nonprovisional application resulting from the conversion of a Provisional application will be measured from the original filing date of the Provisional application. By filing a Provisional application first, and then filing a corresponding Nonprovisional application that references the Provisional application within the 12-month Provisional application pendency period, a patent term endpoint may be extended by as much as 12 months.

Provisional Application for Patent Filing Date Requirements

The Provisional application must name all of the inventor(s). In view of the one-year grace period provided by 35 U.S.C. 102(b)(1) in conjunction with 35 U.S.C. 102(a)(1), a Provisional application can be filed up to 12 months following an inventor's disclosure of the invention. (Such a pre-filing disclosure, although protected in the United States, may preclude patenting in foreign countries.) A public disclosure (e.g., publication, public use, offer for sale) more than one year before the Provisional application filing date would preclude patenting in the United States. Keep in mind that a publication, use, sale, or other activity only has to be made available to the public to qualify as a public disclosure.

A filing date will be accorded to a Provisional application only when it contains a written description of the invention, complying with all requirements of 35 U.S.C. §112(a).

Although the application will be accorded a filing date regardless of whether any drawings are submitted, applicants are advised to file with the application any drawings necessary for the understanding of the invention, complying with 35 U.S.C. 113. A drawing necessary to understand the invention cannot be introduced into an application after the filing date because of the prohibition against new matter. Further, 37 CFR 1.53(c) prohibits amendments from being filed in Provisional applications which are not required to comply with the patent statute and all applicable regulations.

To be complete, a Provisional application **must also** include the filing fee as set forth in 37 CFR 1.16(d) and a cover sheet* identifying:

- the application as a provisional application for patent;
- the name(s) of all inventors;
- inventor residence(s);
- title of the invention;
- name and registration number of attorney or agent and docket number (if applicable);
- correspondence address; and
- any US Government agency that has a property interest in the application.

* A cover sheet, form PTO/SB/16, pages 1 and 2, is included in Appendix C3 that follows and, is available at www.uspto.gov/forms/index.jsp.

Appendix C3

USPTO Forms

The forms in the following pages are updated by the USPTO over time. Therefore, when you are ready to use them, go to the USPTO website and download the most current version of the forms.

1. Form SB0125 Application for Customer Number
2. Form SB0015A Application for Micro Entity Status
3. Form SB0016 Provisional Application Cover Sheet

PTO/SB/125A (11-08)
Approved for use through 11/30/2011. OMB 0651-0035
U.S. Patent and Trademark Office, U.S. DEPARTMENT OF COMMERCE
Under the Paperwork Reduction Act of 1995, no persons are required to respond to a collection of information unless it displays a valid OMB control number.

Request for Customer Number	**Address to:** Mail Stop CN Commissioner for Patents P.O. Box 1450 Alexandria, VA 22313-1450

☐ Although the Requester acknowledges that Internet communications are not secure, the Requester hereby authorizes the USPTO to send the assigned customer number by e-mail to the email address listed below.

To the Commissioner for Patents:
Please assign a Customer Number to the address indicated below:

Firm or Individual Name	
Address	

City		State		Zip	
Country					
Telephone		Email			

Please associate the following practitioner registration number(s) with the Customer Number assigned to the address cited above.

☐ Additional practitioner registration numbers are listed on supplemental sheet(s) attached hereto.

Request Submitted by:

Firm Name (if applicable)			
Signature			
Name of person submitting request		Date	
Registration Number, if applicable		Telephone Number	

This collection of information is required by 37 CFR 1.33. The information is required to obtain or retain a benefit by the public which is to file (and by the USPTO to process) an application. Confidentiality is governed by 35 U.S.C. 122 and 37 CFR 1.11 and 1.14. This collection is estimated to take 12 minutes to complete, including gathering, preparing, and submitting the completed application form to the USPTO. Time will vary depending upon the individual case. Any comments on the amount of time you require to complete this form and/or suggestions for reducing this burden should be sent to the Chief Information Officer, U.S. Patent and Trademark Office, U.S. Department of Commerce, P.O. Box 1450, Alexandria, VA 22313-1450. DO NOT SEND FEES OR COMPLETED FORMS TO THIS ADDRESS. **SEND TO: Mail Stop CN, Commissioner for Patents, P.O. Box 1450, Alexandria, VA 22313-1450.**

If you need assistance in completing the form, call 1-800-PTO-9199 (1-800-786-9199) and select option 2.

Doc Code: MES.GIB
Document Description: Certification of Micro Entity Status (Gross Income Basis)

PTO/SB/15A (07-14)

CERTIFICATION OF MICRO ENTITY STATUS
(GROSS INCOME BASIS)

Application Number or Control Number (if applicable):	Patent Number (if applicable):
First Named Inventor:	Title of Invention:

The applicant hereby certifies the following—

(1) **SMALL ENTITY REQUIREMENT** – The applicant qualifies as a small entity as defined in 37 CFR 1.27.

(2) **APPLICATION FILING LIMIT** – Neither the applicant nor the inventor nor a joint inventor has been named as the inventor or a joint inventor on more than four previously filed U.S. patent applications, excluding provisional applications and international applications under the Patent Cooperation Treaty (PCT) for which the basic national fee under 37 CFR 1.492(a) was not paid, and also excluding patent applications for which the applicant has assigned all ownership rights, or is obligated to assign all ownership rights, as a result of the applicant's previous employment.

(3) **GROSS INCOME LIMIT ON APPLICANTS AND INVENTORS** – Neither the applicant nor the inventor nor a joint inventor, in the calendar year preceding the calendar year in which the applicable fee is being paid, had a gross income, as defined in section 61(a) of the Internal Revenue Code of 1986 (26 U.S.C. 61(a)), exceeding the "Maximum Qualifying Gross Income" reported on the USPTO Web site at http://www.uspto.gov/patents/law/micro_entity.jsp which is equal to three times the median household income for that preceding calendar year, as most recently reported by the Bureau of the Census.

(4) **GROSS INCOME LIMIT ON PARTIES WITH AN "OWNERSHIP INTEREST"** – Neither the applicant nor the inventor nor a joint inventor has assigned, granted, or conveyed, nor is under an obligation by contract or law to assign, grant, or convey, a license or other ownership interest in the application concerned to an entity that, in the calendar year preceding the calendar year in which the applicable fee is being paid, had a gross income, as defined in section 61(a) of the Internal Revenue Code of 1986, exceeding the "Maximum Qualifying Gross Income" reported on the USPTO Web site at http://www.uspto.gov/patents/law/micro_entity.jsp which is equal to three times the median household income for that preceding calendar year, as most recently reported by the Bureau of the Census.

SIGNATURE by an <u>authorized party</u> set forth in 37 CFR 1.33(b)			
Signature			
Name			
Date		Telephone		Registration No.	

☐	There is more than one inventor and I am one of the inventors who are jointly identified as the applicant. The required additional certification form(s) signed by the other joint inventor(s) are included with this form.

PTO/SB/16 (03-13)
Approved for use through 01/31/2014. OMB 0651-0032
U.S. Patent and Trademark Office; U.S. DEPARTMENT OF COMMERCE
Under the Paperwork Reduction Act of 1995 no persons are required to respond to a collection of information unless it displays a valid OMB control number

PROVISIONAL APPLICATION FOR PATENT COVER SHEET – Page 1 of 2

This is a request for filing a PROVISIONAL APPLICATION FOR PATENT under 37 CFR 1.53(c).

Express Mail Label No. _____

INVENTOR(S)		
Given Name (first and middle [if any])	Family Name or Surname	Residence (City and either State or Foreign Country)

Additional inventors are being named on the _____ separately numbered sheets attached hereto.

TITLE OF THE INVENTION (500 characters max):

Direct all correspondence to: **CORRESPONDENCE ADDRESS**

☐ The address corresponding to Customer Number:

OR

☐ Firm or Individual Name

Address

City	State	Zip
Country	Telephone	Email

ENCLOSED APPLICATION PARTS (*check all that apply*)

☐ Application Data Sheet. See 37 CFR 1.76.   ☐ CD(s), Number of CDs _____

☐ Drawing(s)   *Number of Sheets* _____   ☐ Other (specify) _____

☐ Specification (e.g., description of the invention)   *Number of Pages* _____

Fees Due: Filing Fee of $260 ($130 for small entity) ($65 for micro entity). If the specification and drawings exceed 100 sheets of paper, an application size fee is also due, which is $400 ($200 for small entity) ($100 for micro entity) for each additional 50 sheets or fraction thereof. See 35 U.S.C. 41(a)(1)(G) and 37 CFR 1.16(s).

METHOD OF PAYMENT OF THE FILING FEE AND APPLICATION SIZE FEE FOR THIS PROVISIONAL APPLICATION FOR PATENT

☐ Applicant asserts small entity status. See 37 CFR 1.27.

☐ Applicant certifies micro entity status. See 37 CFR 1.29.
Applicant must attach form PTO/SB/15A or B or equivalent.

☐ A check or money order made payable to the *Director of the United States Patent and Trademark Office* is enclosed to cover the filing fee and application size fee (if applicable).   **TOTAL FEE AMOUNT ($)**

☐ Payment by credit card. Form PTO-2038 is attached.

☐ The Director is hereby authorized to charge the filing fee and application size fee (if applicable) or credit any overpayment to Deposit Account Number: _____ .

USE ONLY FOR FILING A PROVISIONAL APPLICATION FOR PATENT

This collection of information is required by 37 CFR 1.51. The information is required to obtain or retain a benefit by the public which is to file (and by the USPTO to process) an application. Confidentiality is governed by 35 U.S.C. 122 and 37 CFR 1.11 and 1.14. This collection is estimated to take 10 hours to complete, including gathering, preparing, and submitting the completed application form to the USPTO. Time will vary depending upon the individual case. Any comments on the amount of time you require to complete this form and/or suggestions for reducing this burden, should be sent to the Chief Information Officer, U.S. Patent and Trademark Office, U.S. Department of Commerce, P.O. Box 1450, Alexandria, VA 22313-1450. DO NOT SEND FEES OR COMPLETED FORMS TO THIS ADDRESS. SEND TO: Commissioner for Patents, P.O. Box 1450, Alexandria, VA 22313-1450.

If you need assistance in completing the form, call 1-800-PTO-9199 and select option 2.

PTO/SB/16 (03-13)
Approved for use through 01/31/2014. OMB 0651-0032
U.S. Patent and Trademark Office; U.S. DEPARTMENT OF COMMERCE
Under the Paperwork Reduction Act of 1995 no persons are required to respond to a collection of information unless it displays a valid OMB control number

PROVISIONAL APPLICATION FOR PATENT COVER SHEET – Page 2 of 2

The invention was made by an agency of the United States Government or under a contract with an agency of the United States Government.

[ ] No.

[ ] Yes, the invention was made by an agency of the U.S. Government. The U.S. Government agency name is: _____
_____

[ ] Yes, the invention was made under a contract with an agency of the U.S. Government. The name of the U.S. Government agency and Government contract number are: _____
_____

WARNING:

Petitioner/applicant is cautioned to avoid submitting personal information in documents filed in a patent application that may contribute to identity theft. Personal information such as social security numbers, bank account numbers, or credit card numbers (other than a check or credit card authorization form PTO-2038 submitted for payment purposes) is never required by the USPTO to support a petition or an application. If this type of personal information is included in documents submitted to the USPTO, petitioners/applicants should consider redacting such personal information from the documents before submitting them to the USPTO. Petitioner/applicant is advised that the record of a patent application is available to the public after publication of the application (unless a non-publication request in compliance with 37 CFR 1.213(a) is made in the application) or issuance of a patent. Furthermore, the record from an abandoned application may also be available to the public if the application is referenced in a published application or an issued patent (see 37 CFR 1.14). Checks and credit card authorization forms PTO-2038 submitted for payment purposes are not retained in the application file and therefore are not publicly available.

SIGNATURE _____ DATE _____

TYPED OR PRINTED NAME _____ REGISTRATION NO. _____
(*if appropriate*)

TELEPHONE _____ DOCKET NUMBER _____

References and Selected Bibliography

ABC TV (USA) (2012). Shark Tank show (Rick Hopper) Nov. 30, Season 3, Episode 302; http://sharktanksuccess.blogspot.com/2012/11/readerest-specsecure.html

Bilton, N. (2014). *Reclaiming Our (Real) Lives From Social Media.* Retrieved from The New York Times, July 16: http://www.nytimes.com/2014/07/17/fashion/reclaiming-our-real-lives-from-social-media.html?_r=0

Biz Info Library, http://www.bizinfolibrary.org/ (accessed January 2016)

Blanchard, K. (2007). *Leading at a Higher Level.* Upper Saddle River, NJ: Pearson.

Blank, Steve and Bob Dorf. (2012). *The Startup Owner's Manual: The Step-by-Step Guide for Building a Great Company,* Pescadero, CA: K and S Ranch Publishers, Inc.

Bplans.com, http://www.bplans.com/sample_business_plans.cfm and also see, http://www.bplans.com/samples/sba.cfm/ (accessed January 2016)

Brandt, L. and Rawski, T. G. (2008). *China's Great Transformation,* Cambridge, UK: Cambridge University Press.

Business plan competitions, http://www.bizplancompetitions.com/ (accessed January 2016)

Center for business planning, http://www.bizplancompetitions.com/ (accessed January 2016)

Christensen, C. M. (1997). *The Innovator's Dilemma,* Boston, MA: Harvard Business School Press.

Christensen, C. M., & Raynor, M. E. (2003). *The Innovator's Solution,* Boston, MA: HBS Press.

Davis, M. (2003). What can we learn by looking for the first code of professional ethics? *Theoretical Medicine and Bioethics,* 24(5): 433–454.

Dyson, J. (2003). *Sir James Dyson, Against the Odds: An autobiography* (2nd ed.), New York: Texere, LLC.

Entrepreneur.com, http://www.entrepreneur.com/ (accessed January 2016)

Eureka Ranch, http://www.eurekaranch.com/ (accessed, April 2015).

Fletcher, N. (2014). *Apple Becomes First Company Worth $700bn. The Guardian*: http://www.theguardian.com/technology/2014/nov/25/apple-first-company-worth-700bn-iphone (accessed November 25, 2014).

Gage, D. (2012). The venture capital secret: 3 out of 4 startups fail. *The Wall Street Journal,* Sept. 19.

Goleman, D., Boyatzis, R. E., & McKee, A. (2002). *Primal Leadership: Realizing the Power of Emotional Intelligence.* Boston, MA: Harvard Business School Press.

Grove, A. (1996). *Only the Paranoid Survive: How to Exploit the Crisis Point that Challenge Every Company.* New York: Doubleday.

GrowAmerica, 10 simple product ideas that made billions, http://learn.growam.com/10-simple-product-ideas/ (accessed July 2013).

Heffernan, M. (2009). James Dyson on creating a vacuum that actually, well, sucks. *Reader's Digest*, February, http://www.rd.com/advice/work-career/james-dyson-on-creating-a-vacuum-that-actually-well-sucks/ (accessed January 2016)

Hickman, G. R. (Ed.). (2010). *Leading Organizations: Perspectives for a New Era* (2nd ed.). Thousand Oaks, CA: Sage.

Hölzl, W. (2010). The economics of entrepreneurship policy: Introduction to the special issue. *Journal of Industry, Competition and Trade*, 10, 187–197.

Isaacson, W. (2011). *Steve Jobs*. New York: Simon and Schuster.

Kastelle. T. (2012). Eight models of business models and why they are important. Innovation for Growth (Blog), 16 January. http://timkastelle.org/blog/2012/01/eight-models-of-business-models-why-theyre-important/ (accessed June 14, 2013).

Kawasaki, G. (Blog), http://blog.guykawasaki.com/#axzz21YXUifxq (accessed January 2016)

Keeney, R. L. (1996). *Value-Focused Thinking: A Path to Creative Decision Making*. Boston, MA: Harvard University Press.

Kim, L. (2015). Top 10 business plan templates you can download free. *Inc.*, June 11. http://www.inc.com/larry-kim/top-10-business-plan-templates-you-can-download-free.html

Kouzes, J., & Posner, B. (2007). *The Leadership Challenge* (4th ed.). San Francisco, CA: Josey-Bass.

Kwoh, L. (2012). You call that innovation? *The Wall Street Journal*, www.WSJ.com May 23.

Mendelsohn, J. (2012). *Constant Contact*. Retrieved from 6 Things You Should Know about Small Businesses in 2012, August 7: http://blogs.constantcontact.com/small-business-pulse-2012/

Morgan, G. (2006). *Images of Organizations*. Thousand Oaks, CA: Sage.

Morrisette, S., & Schraeder, M. (2007). Affirming entrepreneurship: The best hope for organizations. *Development and Learning in Organizations*, 21(1), 15–17.

Null, C. (2013). *TechHive*. Retrieved from Do social media ads really work? We put them to the test!, March 14: http://www.techhive.com/article/2030740/do-social-media-ads-really-work-we-put-them-to-the-test-.html?page=2

Orsini, L. (2013). How an engineering toy for girls went from Kickstarter to best-seller. *Readwrite*, July 12. http://readwrite.com/2013/07/12/how-an-engineering-toy-for-girls-went-from-kickstarter-to-bestseller

Oster, E. (2014). *Advertising & Branding*. Retrieved from AdWeek, August 6: http://www.adweek.com/news/advertising-branding/apple-hires-musa-tariq-head-social-media-marketing-159302

Osterwalder, A., & Pigneur, Y. (2010). *Business Model Generation*. Hoboken, NJ: John Wiley.

Palo Alto Software, http://www.businessplanpro.com/template_offer_lt/?gclid=COuEyP3MsrECFUEGRQodfAMASQ& (accessed January 2016)

Peel, S., & Inkson, K. (2004). Contracting and careers: Choosing between self and organizational employment. *Career Development International*, 9(6/7), 542–558.

Ries, E. (2011). *The lean Startup: How Today's Entrepreneurs Use Continuous Innovation to Create Radically Successful Businesses*, New York: Crown Business.

Roberts, R., & Wood, W. (2007). *Intellectual Virtues: An Essay in Regulative Epistemology*. Oxford, UK: Oxford University Press.

Rocket Lawyer, https://www.rocketlawyer.com/secure/interview/questions.aspx?document=12449668&utm_source=103&v=3#q1 (accessed January 2016)

Rus, D., van Knippenberg, D., & Wisse, B. (2012). Leader power and self-serving behavior: The moderating role of accountability. *The Leadership Quarterly, 23*(1), 13–26. doi: 10.1016/j.leaqua.2011.11.002.

Said, C. (2014). GoldieBlox Super Bowl ad strives to entice girls to engineering, SFGate website of the San Francisco Chronicle newspaper, Feb. 3. http://www.sfgate.com/business/article/GoldieBlox-Super-Bowl-ad-strives-to-entice-girls-5194465.php (accessed January 2016)

Schwartz, A. (2014). How GoldieBlox went from a scrappy Kickstarter to making important toys for girls. *Fastcompany*, Sept 15. http://www.fastcoexist.com/3035356/how-goldieblox-went-from-a-scrappy-kickstarter-to-making-important-toys-for-girls (accessed January 2016)

Schyns, B., & Hansbrough, T. (Eds.). (2010). *When Leadership Goes Wrong: Destructive Leadership, Mistakes, and Ethical Failures*. Charlotte, NC: Information Age Publishing.

Small Business Administration, http://www.sba.gov/ (accessed January 2016)

Sophia. (2013). *Thought Reach*. Retrieved from 20 Facts About Social Media Marketing for Small Businesses: http://thoughtreach.com/facts-about-social-media-marketing-for-small-businesses/

Swamidass, P. (2010a). Does the patent office snub inventors without an attorney. Inventors Digest, April; http://idmagazine.wpengine.com/articles/does-the-patent-office-snub-inventors-who-file-without-an-attorney/ (accessed January 2016)

Swamidass, P. (2010b). Reforming the USPTO to comply with MPEP section 707.07(J) to give a fair shake to pro se inventor-applicants. *John Marshall Review of Intellectual Property Law*, 9(4), 880–911.

Swamidass, P. (2014). Engineers and scientists: Value creators in the seven-phased model of technological innovation. *Technology and Innovation,* 16 (3-4), 223–232.

Swamidass, P., & Schnittka, C. (Forthcoming). Funding and preparing teachers to meet the needs of US student innovators in the making, in *Technology and Innovation*, the Journal of The National Academy of Inventors.

Ulrich, K., & Eppinger, S. (2015). *Product Design and Development* (6th ed.). New York: McGraw-Hill Education.

Utterback, J. M. (1996). *Mastering the Dynamics of Innovation*. Boston, MA: HBS Press.

USPTO (2014). 16th Annual Independent Inventors Conference, August 15–16. http://www.uspto.gov/about-us/events/16th-annual-independent-inventors-conference (accessed January 2016)

USPTO, Checklist for filing a Nonprovisional Utility Patent Application, http://www.uspto.gov/sites/default/files/inventors/Checklist_for_Filing_a_Nonprovisional_Utility.pdf (accessed January 2016)

USPTO, US Patent Classification System, http://www.uspto.gov/page/examiner-handbook-us-patent-classification-system (accessed January 2016)

USPTO (2015). Patent "Lunch and Learn," Nov. 19, Silicon Valley USPTO, San Jose, CA. http://www.uspto.gov/about-us/events/patent-lunch-and-learn (accessed January 2016)

USPTO, Patent Pro Bono Program, http://www.uspto.gov/patents-getting-started/using-legal-services/pro-bono/inventors (accessed January 2016)

USPTO, Published Complaints, http://www.uspto.gov/patents-getting-started/using-legal-services/scam-prevention/published-complaints (accessed January 2016)

USPTO, Independent inventors by state by year: Utility Patents Report, January 1975–December 2014. http://www.uspto.gov/web/offices/ac/ido/oeip/taf/inv_utl.htm (accessed January 2016)

USPTO, Inventors & Entrepreneurs Resources, http://www.uspto.gov/learning-and-resources/inventors-entrepreneurs-resources (accessed January 2016)

USPTO, *Inventors Eye* Newsletter, http://www.uspto.gov/learning-and-resources/inventors-eye-newsletter (accessed January 2016)

USPTO Provisional Application for Patent: http://www.uspto.gov/patents-getting-started/patent-basics/types-patent-applications/provisional-application-patent (accessed January 2016)

USPTO video on provisional patent application, https://www.youtube.com/watch?v=qIGrAg5AVL8 (accessed January 2016)

USPTO video for submitting a provisional application electronically, https://www.youtube.com/watch?v=B--8oFg88Uw (accessed January 2016)

USPTO, Webinar for Patent Pro Se, Pro Bono, and Law School Clinic Certification Programs. http://www.uspto.gov/sites/default/files/documents/ProBono_ProSe_Law_School_Clinic_Certification_Program.pdf (accessed January 2016)

van Knippenberg, B., & van Knippenberg, D. (2005). Leader self-sacrifice and leadership effectiveness: The moderating role of leader prototypicality. *Journal of Applied Psychology, 90,* 25–37.

Index